AGRICULTURAL MACHINERY MANAGEMENT

AGRICULTURAL MACHINERY MANAGEMENT

N. Selvaganesan
Aayushi Deohans
Priyadarsini Sahoo

KNOWLEDGE
BAKERS

KNOWLEDGE BAKERS

ISBN: 978-93-49964-32-7

Published by : **Knowledge Bakers**
Pune, India
info@knowledgebakers.com

PREFACE

Efficient use and management of agricultural machinery are essential components of modern farming, directly influencing productivity, sustainability, and profitability. As agriculture evolves with technological innovations, the integration of mechanized systems in cultivation, harvesting, processing, and transportation has become increasingly critical. Understanding the principles of machinery selection, operation, maintenance, and economic evaluation empowers farmers, engineers, and agricultural professionals to make informed decisions that enhance field efficiency and reduce operational costs.

This handbook, titled Agricultural Machinery Management, provides a comprehensive overview of the technical and managerial aspects of farm equipment. It draws upon practical experience, research insights, and engineering fundamentals to present a structured approach to machinery management tailored for contemporary agricultural settings.

Covering key topics such as machinery performance metrics, power requirements, equipment calibration, scheduling, maintenance planning, and cost analysis, Agricultural Machinery Management equips readers with the necessary tools to optimize machine usage. Real-world examples, illustrative diagrams, and step-by-step procedures enhance understanding and applicability across various scales of farming operations.

Moreover, the book explores the role of digital tools, automation, and precision agriculture in machinery management, highlighting emerging trends and practices that are reshaping the agricultural landscape. Environmental impacts, energy efficiency, and sustainable machinery practices are also addressed, reflecting the growing importance of responsible resource use in agriculture.

It is our hope that Agricultural Machinery Management will serve as a valuable guide for students, practitioners, and researchers in the agricultural sciences and engineering. Whether in the classroom, workshop, or field, this volume offers insights and practical knowledge to support the effective and sustainable use of agricultural machinery in the years ahead.

N. Selvaganesan
Aayushi Deohans
Priyadarsini Sahoo

CONTENTS

Chapter 1

INTRODUCTION

1.1 OVERVIEW

Agricultural Machine Technician; repairs, services and maintains agricultural machinery. The machinery may be large or small, basic or advanced, and include tractors, forage harvesters, balers, bale wrappers, mowers, combine harvesters, crop sprayers, fertiliser distributors, tillage and slurry handling equipment. According to type and purpose, the machinery may be used primarily or entirely off the road.

For servicing and repair the Agricultural Machine Technician may normally work on the owner's premises and land, which is usually but not always a farm. Some Technicians will be based in a workshop that serves a particular area or type of machine. The Technician will be required to provide both routine and exceptional support to those owning or leasing the machinery. Some machines and units have a largely seasonal use, while others are required for many purposes year-round. Whatever the requirement, for this time-sensitive sector reliability, efficiency and effectiveness are essential.

Agricultural Machine Technician must be a good problem solver with strong diagnostic and analytical skills. He or she must be able to work on both simple and complex machines of a traditional or very modern kind. While some physical strength is required, there is a growing need for electronic skills as new technologies impact upon agricultural methods. Agriculture and agricultural machines are associated with multiple, serious hazards. The Agricultural Machine Technician has an important role to play in keeping him- or herself safe, and in actively supporting the health and safety of others. This is a key requirement in all the environments and conditions that he or she is called to work in and with.

Agricultural Machine Technician generally works in close association with farmers and other machine users for whom delays are a major problem. Speed and efficiency are therefore looked for, together with understanding and empathy. With the dual pressures of the need to produce more food while sustaining the environment, the Agricultural Machine Technician will need to keep abreast with rapid modern developments in farming methods, equipment and machinery. These will impact dramatically on his or her role and skills requirements.

1.2 DUTIES

Agricultural equipment technicians repair, overhaul and maintain agricultural equipment including tractors, tillage equipment, seeding equipment and harvesting equipment.

Agricultural equipment technicians may specialize in service and repair of fuel injection pumps and injectors, engine overhaul, hydraulic systems, power shift transmissions or specific types of equipment (combines, forage harvesters, tractors, etc.).

In general, agricultural equipment technicians:

- ☆ service and repair engines, transmissions, hydraulic systems, and the electrical and electronic systems of agricultural equipment,
- ☆ pinpoint problems by discussing equipment operation with owners, visually checking and listening to machinery in operation, and using testing equipment,
- ☆ take defective units apart and repair or replace broken, worn-out or faulty parts, reassemble, adjust, lubricate and test repaired units,
- ☆ assemble and adjust new agricultural equipment, and
- ☆ learn and apply precision farm technologies such as automated steering systems.
- ☆ Agricultural equipment technicians must keep up-to-date with new developments in the agricultural equipment field.

1.3 SKILLS AND ABILITIES

The work is most rewarding for those who enjoy working with their hands and who are capable of diagnosing a variety of complex mechanical, hydraulic and electrical problems. To be successful in the trade, agricultural equipment technicians need:

- ☆ mechanical ability and good eye-hand coordination,
- ☆ the strength required to lift heavy machine parts,
- ☆ the ability to read and understand technical information,
- ☆ the ability to work independently with little supervision,
- ☆ flexibility in working extra hours when required,
- ☆ some experience in the agriculture industry,
- ☆ math, communication and computer skills,
- ☆ effective interpersonal skills in meeting the challenges of the workplace, and
- ☆ a valid driver's license.

1.4 EMPLOYMENT AND ADVANCEMENT

Most agricultural machine technicians are employed by retail agricultural equipment dealerships. Where a number of technicians are employed by one business, the opportunity exists for advancement to supervisory positions.

Agricultural equipment technicians can move into sales positions or become manufacturers' service representatives. Individuals may move into management positions with the company or purchase an agricultural equipment dealership.

1.5 KNOWLEDGE ABOUT ENERGY AND CONSERVATION

Energy does not come cheaply. In addition to the economic costs, society pays a huge environmental price for its use of nonrenewable energy in damage to the health of its people and to the environment. These impacts arise at every phase of energy production. The most significant impacts arise from extraction and end use.

1.6 NATURAL RESOURCES

Natural resources are anything we take from the earth and use. Examples or natural resources are oil, minerals, water, plants, animals, rocks and the air we breathe. Natural resources can be categorized into three different groups – renewable, nonrenewable, and inexhaustible. Some natural resources will eventually be used up so conservation of these resources is very important. Fossil fuels and non-renewable sources of energy- coal, natural gas, gas generated from landfills crude oil, nuclear power. Many nations count on coal, oil and natural gas to supply most of their energy needs, but reliance on fossil fuels presents a big problem. Fossil fuels are a finite resource. Eventually, the world will run out of fossil fuels, or it will become too expensive to retrieve those that remain. Fossil fuels also cause air, water and soil pollution, and produce greenhouse gases that contribute to global warming. Fuels generally create waste byproducts, some of which can be harmful pollutants.

Natural resources are available to sustain the very complex interaction between living things and non-living things. Humans also benefit immensely from this interaction. All over the world, people consume resources directly or indirectly. Developed countries consume resources more than under-developed countries. In what form do people consume natural resources? The three major forms include Food and drink, Housing and infrastructure, and Mobility. These three make up more than 60% of resource use.

1.6.1 Forms

Food and drink:

This includes agricultural products as well as naturally occurring foods such as game, fish from fresh water and seas, seeds and nuts, medicines, herbs and plants. They also include drinking water, as well as water for sanitation and household use. Think of ceramic plates, silverware (spoons, forks and knives), cans, milk packages, paper and plastic cups — they are all made from raw materials which come from our natural resources.

Mobility:

This includes automobiles, trains, water vessels, airplanes, together with all the fuel that powers them. Can you imagine where all the raw materials used in their production came from?

Housing and infrastructure:

Think about all the houses, public places, roads and constructed objects you have in your city or town. Think about all the energy for heating and cooling that we consume in our homes — can you imagine where all the wood, metals, plastic, stone and other materials came from?

Beyond these three major areas of resource consumption, we consume much more resources from our environment on a daily basis. The role of natural resources in sustaining life on earth is extremely important and we must ensure that we protect the environment and also make it easy for it to replenish itself naturally.

1.6.2 Threats to Natural Resources

A. Overpopulation

This is probably the most significant, single threat that natural resources face. The world's population is increasing at a very fast rate. In the USA, a baby is born every

8 seconds, and a person dies every 13 secondssource1 The increase in populations mean there will be pressure on almost all natural resources. How?

- ☆ Land Use: With more mouths to feed and people to house, more land will need to be cultivated and developed for housing. More farming chemicals will be applied to increase food production. Many forest or vegetative lands will be converted to settlements for people, roads and farms. These have serious repercussions on natural resources.
- ☆ Forests: Demand for wood (timber), food, roads and forest products will be more. People will therefore use more forest resources than they can naturally recover.
- ☆ Fishing: Fresh water and sea food will face problems too as we will continue to depend heavily on them. Bigger fishing companies are going deeper into sea to catch fish in even larger quantities. Some of the fishing methods they use are not sustainable, thereby destroying much more fish and sea creatures in the process.
- ☆ Need for more: Human's demand for a comfortable life means more items (communication, transport, education, entertainment and recreation) will need to be produced. This means more industrial processes and more need for raw materials and natural resources.

B. Climate Change

The alteration in climate patterns as a result of excessive anthropogenic Carbon dioxide is hurting biodiversity and many other abiotic natural resources. Species that have acclimatized to their environments may perish and others will have to move to more favorable conditions to survive.

C. Environmental Pollution

Land, water and air pollution directly affect the health of the environments in which they occur. Pollution affects the chemical make-up of soils, rocks, lands, ocean water, freshwater and underground water, and other natural phenomena. This often has catastrophic consequences.

1.7 KNOWLEDGE ABOUT RENEWABLE ENERGY RESOURCES

Renewable energy resources such as wind, solar and hydropower, offer clean alternatives to fossil fuels. They produce little or no pollution or greenhouse gases, and they will never run out.

1.7.1. Solar Energy

The sun is our most powerful source of energy. Sunlight, or solar energy, can be used for heating, lighting and cooling homes and other buildings, generating electricity, water heating, and a variety of industrial processes. Most forms of renewable energy come either directly or indirectly from the sun. For example, heat from the sun causes the wind to blow, contributes to the growth of trees and other plants that are used for biomass energy, and plays an essential role in the cycle of evaporation and precipitation that makes hydropower possible.

1.7.2. Wind Energy

Wind is the movement of air that occurs when warm air rises and cooler air rushes in to replace it. The energy of the wind has been used for centuries to sail ships and drive windmills that grind grain. Today, wind energy is captured by wind turbines and used to generate electricity.

1.7.3. Hydropower

Water flowing downstream is a powerful force. Water is a renewable resource, constantly recharged by the global cycle of evaporation and precipitation. The heat of the sun causes water in lakes and oceans to evaporate and form clouds. The water then falls back to Earth as rain or snow, and drains into rivers and streams that flow back to the ocean. Flowing water can be used to power water wheels that drive mechanical processes. And captured by turbines and generators, like those housed at many dams around the world, the energy of flowing water can be used to generate electricity.

1.7.4. Biomass Energy

Biomass has been an important source of energy ever since people first began burning wood to cook food and warm themselves against the winter chill. Wood is still the most common source of biomass energy, but other sources of biomass energy include food crops, grasses and other plants, agricultural and forestry waste and residue, organic components from municipal and industrial wastes, even methane gas harvested from community landfills. Biomass can be used to produce electricity and as fuel for transportation, or to manufacture products that would otherwise require the use of non-renewable fossil fuels.

1.7.5. Hydrogen

Hydrogen has tremendous potential as a fuel and energy source, but the technology needed to realize that potential is still in the early stages. Hydrogen is the most common element on Earth—for example, water is two-thirds hydrogen—but in nature it is always found in combination with other elements. Once separated from other elements, hydrogen can be used to power vehicles, replace natural gas for heating and cooking, and to generate electricity.

1.7.6. Geothermal Energy

The heat inside the Earth produces steam and hot water that can be used to power generators and produce electricity, or for other applications such as home heating and power generation for industry. Geothermal energy can be drawn from deep underground reservoirs by drilling, or from other geothermal reservoirs closer to the surface.

1.7.7. Ocean Energy

The ocean provides several forms of renewable energy, and each one is driven by different forces. Energy from ocean waves and tides can be harnessed to generate electricity, and ocean thermal energy—from the heat stored in sea water—can also be converted to electricity. Using current technologies, most ocean energy is not cost-effective compared to other renewable energy sources, but the ocean remains and important potential energy source for the future.

1.8 KNOWLEDGE ABOUT NON-RENEWABLE RESOURCES

A nonrenewable resource is a resource of economic value that cannot be readily replaced by natural means on a level equal to its consumption. Most fossil fuels, such as oil, natural gas and coal are considered nonrenewable resources in that their use is not sustainable because their formation takes billions of years.

1.8.1 Types of Nonrenewable Resources

Fuels created from nonrenewable resources are still the primary source of all the power generated in the world, due to their affordability and high energy content. Usually, nonrenewable resources are formed from organic carbon material which, over the course of billions of years, is heated and compressed enough to change form into crude oil and natural gas. Crude oil and natural gas are then used to manufacture everyday forms of fuel, including gasoline, diesel fuel, heating oil and propane.

The term nonrenewable resource also refers to minerals and metals from the earth, such as gold, silver, and iron, which are similarly formed as a long-term result of geological processes such as plate tectonics. These resources are often costly to mine, as they are usually deep within the Earth's crust, but they are much more abundant than fossil fuels. Some types of groundwater are considered to be a nonrenewable resource, if the aquifer is unable to be replenished at the same rate at which it's drained. Also, nuclear materials such as uranium are nonrenewable resources.

1.8.1.1 Fossil Fuels

Fossil fuels are derived from organic matter which has been trapped between layers of sediments within the Earth for millions of years. The organic matter, typically plants, have decomposed and compressed over time, leaving what are known as fossil fuel deposits. These deposits, and the materials produced from them, tend to be highly combustible, making them an ideal energy source. They are difficult to obtain as they are typically retrieved through drilling or mining, but fossil fuels are worth the effort for the sheer amount of energy they produce.

The fossil fuels that we use to produce electricity and to drive our cars have some big disadvantages. They produce carbon dioxide, which is a greenhouse gas, and they produce other pollutants too. What's more, once they have all been used there will be no other similar fuels that we can use to replace them- unless we think about the problems and do something about it.

Plants may be one answer to the problem of fuels. For thousands of years people have burned wood to keep themselves warm. Obviously we cannot use wood as a fuel for cars, but there are two ways that plants may be able to keep us on the road.

1. We can use plants that make sugar to produce ethanol by fermenting the sugar using yeast. We can than add the ethanol to petrol, making **gasohol.** Not only does this reduce the amount of oil needed, it also produces less pollution because gasohol burns more cleanly than pure petrol.
2. Another new fuel is **biodiesel.** Some plants (e.g. oilseed) produce oils which can be used in diesel engines. We hardly need to make any changes to the engine to do this, and the biodiesel burns very cleanly, like gasohol. These bio-fuels also help tackle global warming. That's because the plants take in carbon dioxide gas during photosynthesis. They still give off carbon dioxide when we burn the bio-fuel- but overall they make little contribution to the greenhouse effect compared with burning fossil fuels.

Another way that we could reduce the amount of fuels we use is to replace them with something else- and rubbish seems a good answer. By **burning rubbish** we could produce some of the energy we need to heat our homes, and we could get rid of a big problem too.

Getting rid of all our rubbish usually means burying it in holes in the ground. This is not a good solution since it is messy, smelly and produces pollution.

But producing energy from rubbish is not straightforward. Unless the incinerator is run very carefully, dangerous chemicals called dioxins may be produced when the rubbish burns. Although no-one is exactly certain what dioxins do, many people think that they may cause cancer, and that they may damage us in other ways too. So there are arguments on both sides about the benefits of building incinerators.

Burning of fuels

Burning fuels produce carbon dioxide, (which might be a greenhouse gas and cause global warming), particulates= unburnt hydrocarbons, (which travel into the upper atmosphere and cause global dimming), carbon monoxide (serious for people with heart problems), sulfur dioxide and nitrogen oxides (which can fall as acid rain). We can reduce the effect of burning fossil fuels by removing the pollutants from the gases that are produced when we burn fuels. For some time the exhaust systems of cars have been fitted with catalytic converters. In power stations, sulfur dioxide is removed from the flue gases by reacting it with quicklime. This is called flue gas desulfurisation.

Formation of Fossil Fuels

Contrary to what many people believe, fossil fuels are not the remains of dead dinosaurs. In fact, most of the fossil fuels we find today were formed millions of years before the first dinosaurs. Fossil fuels, however, were once alive as they were formed from prehistoric plants and animals that lived hundreds of millions of years ago.

Think about what the Earth must have looked like 300 million years or so ago. The land masses we live on today were just forming. There were swamps and bogs everywhere. The climate was warmer. Ancient trees and plants grew everywhere. Strange looking animals walked on the land, and just as weird looking fish swam in the rivers and seas. Tiny one-celled organisms called protoplankton floated in the ocean.

When these ancient living things died, they decomposed and became buried under layers and layers of mud, rock, and sand. Eventually, hundreds and sometimes thousands of feet of earth covered them. In some areas, the decomposing materials were covered by ancient seas, then the seas dried up and receded.

During the millions of years that passed, the dead plants and animals slowly decomposed into organic materials and formed fossil fuels. Different types of fossil fuels were formed depending on what combination of animal and plant debris was present, how long the material was buried, and what conditions of temperature and pressure existed when they were decomposing.

For example, oil and natural gas were created from organisms that lived in the water and were buried under ocean or river sediments. Long after the great prehistoric seas and rivers vanished, heat, pressure and bacteria combined to compress and „cook" the organic material under layers of silt. In most areas, a thick liquid called oil formed first, but in deeper, hot regions underground, the cooking process continued until natural gas was formed. Over time, some of this oil and natural gas began working its way upward through the earth's crust until they ran into rock formations called

„caprocks“ that are dense enough to prevent them from seeping to the surface. It is from under these caprocks that most oil and natural gas is produced today.

The same types of forces also created coal, but there are a few differences. Coal formed from the dead remains of trees, ferns and other plants that lived 300 to 400 million years ago. In some areas, such as portions of what-is-now, coal was formed from swamps covered by sea water. The sea water contained a large amount of sulfur, and as the seas dried up, the sulfur was left behind in the coal. Today, scientists are working on ways to take the sulfur out of coal because when coal burns, the sulfur can become an air pollutant.

Some coal deposits, however, were formed from freshwater swamps which had very little sulfur in them.

1.8.1.2 Crude Oil

Oil is extracted from deep wells on the seafloor and on land; it is often found in association with natural gas. After it is extracted, crude oil is heated and distilled, a process that separates the components of oil, which produces useful fuel and nonfuel by-products. The major impacts of the oil energy system come from oil spills and from combustion of oil and its by-products.

1.8.1.3 Natural Gas

Natural gas is a combustible gas extracted from deep onshore and offshore wells. Natural gas is used to heat homes, cook food, and heat water. It is also used by factories and some power plants. It is easy to transport within countries or from one country to its terrestrial neighbors. Natural gas is a clean fuel compared to oil, oil by-products, and coal.

1.8.1.4 Coal

Coal is the most abundant fossil fuel, but its extraction and use create enormous social, economic, and environmental costs. Coal is removed by surface and underground mines, both of which create many environmental impacts. Surface mining is especially damaging to the environment; proper controls can greatly reduce impacts, though, and mine reclamation can help companies to return lands to their original use.

Underground coal mines present health and safety hazards for miners. Many miners have been killed or injured by accidents. Many others suffer from black lung disease, a crippling degenerative lung disease, as a result of exposure to coal dust. One of the biggest environmental problems from coal mining results from the release of sulfuric acid from abandoned underground mines, which poisons thousands of miles of streams.

Electric power plants are the major consumer of coal. Although pollution controls reduce the amount entering the atmosphere, the large number of power plants and the enormous quantities of coal burned in them result in the production of massive amounts of air pollution. Several air pollutants contribute to two of the world's most pressing environmental problems - acid deposition and global climate change (global warming). Moreover, pollutants captured by pollution control devices end up as solid waste, much of which is buried in landfills.

Coal Gasification and Liquefaction

Coal can be converted to gaseous and liquid fuels to replace oil and natural gas, but the processes are energy-intensive and produce much pollution.

1.8.1.5 Oil Shale

Oil shale is a sedimentary rock containing an organic material (kerogen) that can be extracted from the rock by heating. In a liquid state, this thick, oily substance, called shale oil, can be refined to make gasoline and a host of other chemical by-products. Although it is abundant, oil shale is economically and environmentally costly to produce and is currently an insignificant source of fuel.

1.8.1.6 Tar Sands

Tar sands are sand deposits impregnated with a petroleum like substance known as bitumen. Although there are large deposits, they are economically and environmentally costly to extract.

Chapter 2

AGRICULTURE TOOLS

2.1 INTRODUCTION

There are a number of appropriate technology principles that specifically concern agricultural tools. Such tools needs to be produced in the country, and can be repairable at local level. With much of agriculture characterized by short intense periods of activity, farmers cannot afford delays caused by equipment failures. There are certain general principles related to appropriate agricultural tools, some of which, goes beyond the general criteria for appropriate technology which includes:

a) adapted to allow efficient and speedy work with the minimum of fatigue;

b) not injurious to man or animal;

c) of simple design, so that they can be made locally;

d) light in weight, for easy transportation (there are also considerable advantages when threshers, winnowers, and machines such as coffee hullers can be easily moved to where they are needed;

e) ready for immediate use without loss of time for preparatory adjustments;

f) made of easily available materials.

Appropriate agricultural tools and equipment contribute to the broad objective of increasing the viability of agriculture. Where small farmers are currently employing traditional technologies that are inefficient, they often cannot improve this technology because of the leap in scale and capital cost to commercially available equipment. It is therefore the goal of intermediate technology proponents to help fill this gap with good quality tools and equipment that are affordable and suited to the scale of operations of the small farmers. There is a tendency for equipment development and commercial firms to concentrate their energies on tools that are affordable only to the wealthier farmers.

There are certain contributing factors which includes inappropriate application of industrialized, extensive farming strategies to small intensive farming communities, and failure to include small farmer in process of identifying helpful new technologies that can truly fit into the existing farming system. The result is usually either outright failure of innovations to attract interest or the consolidation of landholdings by wealthier farmers taking advantage of the technology newly available. The position of tenant farmer may become worse, and that of small farmer in general is not improved. Appropriate technology advocates must be careful to avoid repeating these mistakes. The degree of concentration of land ownership is a key factor in determining if there are opportunities available for appropriate technology strategies in a community.

Productivity per acre or hectare may in fact decline if large tools require extra space to maneuver and wide lanes to drive or roll over. The appropriate tools under such circumstances, even if supported by unlimited resources, would be very different, where the amount of cultivated land per capita is relatively large. From the national perspective, support for communities of small farms should bring significant benefits. Whereas it has been widely assumed that only the large farm could efficiently increase national food production in the struggle against hunger, mounting evidence from many countries indicates that the small farm has higher yields per acre and plays a crucial role in the distribution of food.

People interested in improving local agricultural equipment should be looking for technologies that accomplish one or more of the following:

1. Remove labor bottlenecks in the agricultural calendar that are limiting production (e.g., short periods of time when all available labor is fully employed, such as during planting or harvesting).
2. Replace or speed up activities that are extremely inefficient in the use of time (e.g., traditional hand-milling). This can free time for more productive activities.
3. Increase the productivity of land (e.g., with irrigation weeding, natural fertilizers) The effectiveness of efforts to create relevant new tools can be increased by concentrating on some key agricultural activities.

Traditional trucks and tractors and the various implements they pull have saved countless hours of labor and will continue to do so. Now, however, the smaller, lighter versions have become affordable even for hobby farmers.

2.2 DIFFERENT TYPES OF AGRICULTURAL TOOLS

2.2.1 Heavy Equipment

This category holds the most extensive array of traditional agricultural equipment: tractors, manure spreaders, fertilizer and insecticide sprayers, plows, cultivators, harrows, planters, harvesters, balers, wagons and trucks weighing more than one ton, with six- to-18 wheels.

Larger concerns can't afford to farm without these big-ticket items, while smaller farms and ranches either band together to buy or lease tractors, trucks and the implements they pull. The tractor pulls the plow through each field in spring or fall, followed by the disc or harrow. Once the planter trails through the pasture, the wait begins. At the end of summer, the harvester takes grain and leaves the hay behind, or takes corn and leaves silage. After a few days of drying time, a baler turns all that chaff into round or square bales. Large trucks transport the entire harvest from field to packinghouse to store.

2.2.2 Light Equipment

Hobby farmers and homesteaders have a tractor-drawn or a walk-behind tiller at their disposal. With the right lawn tractor, a hobby farmer can plow everything from a kitchen garden to one of the long, narrow land parcels known as "bowling-alley" lots. Snowmobiles and 4-wheelers help ranchers and other livestock farmers haul feed and round up stock, run fencelines or harass predators.

2.2.3 Hand Tools

Hand tools are tools that are powered manually. Hand tools include anything from axes to wrenches. The greatest hazards posed by hand tools result from misuse and improper maintenance.

Hazards

Some examples include the following:

- ☆ If a chisel is used as a screwdriver, the tip of the chisel may break and fly off, hitting the user or other employees.

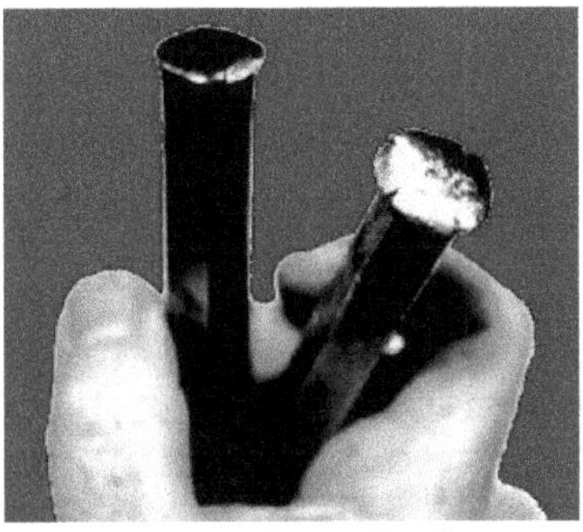

Fig Breaking of tool

☆ If a wooden handle on a tool, such as a hammer or an axe, is loose, splintered, or cracked, the head of the tool may fly off and strike the user or other employees.

Fig Crack in widen piece

☆ If the jaws of a wrench are sprung, the wrench might slip.

☆ If impact tools such as chisels, wedges, or drift pins have mushroomed heads, the heads might shatter on impact, sending sharp fragments flying toward the user or other employees.

Fig Working with hand tools

☆ The employer is responsible for the safe condition of tools and equipment used by employees. Employers shall not issue or permit the use of unsafe hand tools. Employees should be trained in the proper use and handling of tools and equipment.

☆ Employees, when using saw blades, knives, or other tools, should direct the tools away from aisle areas and away from other employees working in close proximity. Knives and scissors must be sharp; dull tools can cause more hazards than sharp ones. Cracked saw blades must be removed from service.

- ☆ Wrenches must not be used when jaws are sprung to the point that slippage occurs. Impact tools such as drift pins, wedges, and chisels must be kept free of mushroomed heads. The wooden handles of tools must not be splintered.
- ☆ Iron or steel hand tools may produce sparks that can be an ignition source around flammable substances. Where this hazard exists, spark-resistant tools made of non-ferrous materials should be used where flammable gases, highly volatile liquids, and other explosive substances are stored or used.

2.2.4 Power Tools

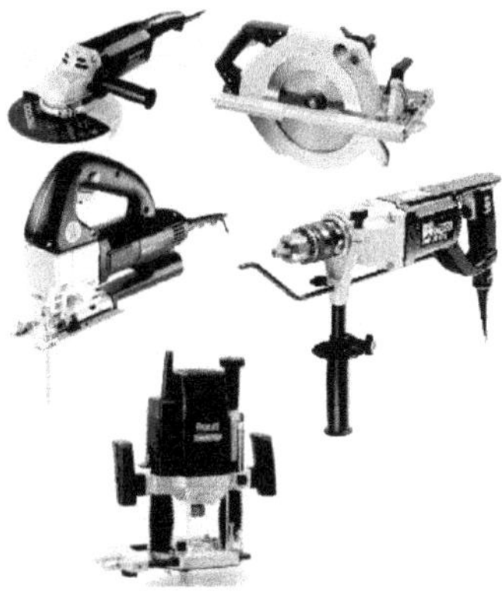

Fig Power Tools

Dangers of Power Tools

Appropriate personal protective equipment such as safety goggles and gloves must be worn to protect against hazards that may be encountered while using hand tools. Workplace floors shall be kept as clean and dry as possible to prevent accidental slips with or around dangerous hand tools.

Power tools must be fitted with guards and safety switches; they are extremely hazardous when used improperly. The types of power tools are determined by their power source: electric, pneumatic, liquid fuel, hydraulic, and powder-actuated.

To prevent hazards associated with the use of power tools, workers should observe the following general precautions:

- ☆ Never carry a tool by the cord or hose.
- ☆ Never yank the cord or the hose to disconnect it from the receptacle.
- ☆ Keep cords and hoses away from heat, oil, and sharp edges.
- ☆ Disconnect tools when not using them, before servicing and cleaning them, and when changing accessories such as blades, bits, and cutters.
- ☆ Keep all people not involved with the work at a safe distance from the work area.

- ☆ Secure work with clamps or a vise, freeing both hands to operate the tool.
- ☆ Avoid accidental starting. Do not hold fingers on the switch button while carrying a plugged-in tool.
- ☆ Maintain tools with care; keep them sharp and clean for best performance.
- ☆ Follow instructions in the user's manual for lubricating and changing accessories.
- ☆ Be sure to keep good footing and maintain good balance when operating power tools.
- ☆ Wear proper apparel for the task. Loose clothing, ties, or jewelry can become caught in moving parts.
- ☆ Remove all damaged portable electric tools from use and tag them: "Do Not Use."

2.3 WORKING WITH DIFFERENT TOOLS

2.3.1 Electric Tools

Technicians using electric tools must be aware of several dangers. Among the most serious hazards are electrical burns and shocks.

- ☆ Electrical shocks, which can lead to injuries such as heart failure and burns, are among the major hazards associated with electric powered tools. Under certain conditions, even a small amount of electric current can result in fibrillation of the heart and death. An electric shock also can cause the user to fall off a ladder or other elevated work surface and be injured due to the fall.
- ☆ To protect the user from shock and burns, electric tools must have a three-wire cord with a ground and be plugged into a grounded receptacle, be double insulated, or be powered by a low voltage isolation transformer. Three-wire cords contain two current carrying conductors and a grounding conductor. Any time an adapter is used to accommodate a two-hole receptacle, the adapter wire must be attached to a known ground. The third prong must never be removed from the plug.

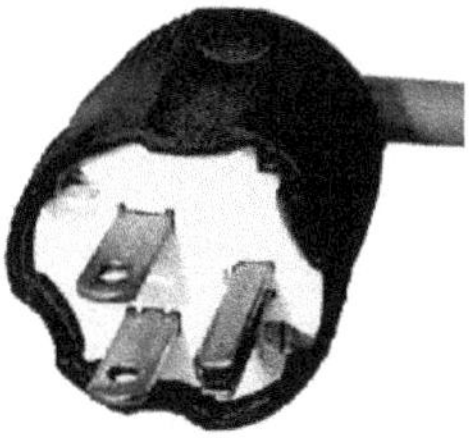

Fig Three Wire Plug

Double-insulated tools are available that provide protection against electrical shock without third-wire grounding. On double insulated tools, an internal layer of protective insulation completely isolates the external housing of the tool. The following general practices should be followed when using electric tools:

- ☆ Operate electric tools within their design limitations.
- ☆ Use gloves and appropriate safety footwear when using electric tools.
- ☆ Store electric tools in a dry place when not in use.
- ☆ Do not use electric tools in damp or wet locations unless they are approved for that purpose.
- ☆ Keep work areas well lighted when operating electric tools.
- ☆ Ensure that cords from electric tools do not present a tripping hazard.

2.3.2 Portable Abrasive Wheel Tools

Portable abrasive grinding, cutting, polishing, and wire buffing wheels create special safety problems because they may throw off flying fragments. Abrasive wheel tools must be equipped with guards that:

1. cover the spindle end, nut, and flange projections;
2. maintain proper alignment with the wheel;
3. do not exceed the strength of the fastenings.

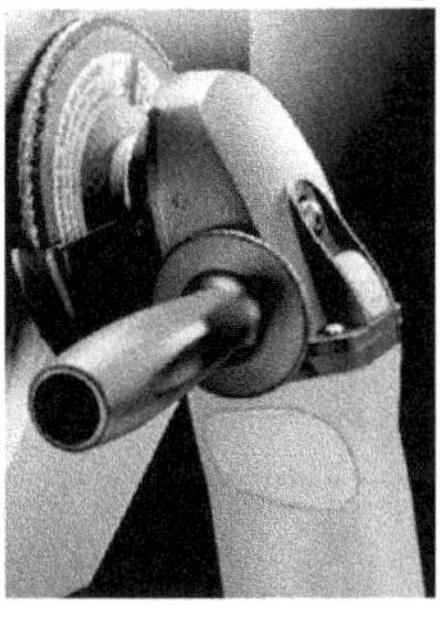

Fig Abrasive wheel tool

Before an abrasive wheel is mounted, it must be inspected closely for damage and should be sound- or ring-tested to ensure that it is free from cracks or defects. To test, wheels should be tapped gently with a light, non-metallic instrument. If the wheels found cracked or dead, they must not be used because they could fly apart in operation. A stable and undamaged wheel, when tapped, will give a clear metallic tone or "ring."

Fig Inspection of tool

To prevent an abrasive wheel from cracking, it must fit freely on the spindle. The spindle nut must be tightened enough to hold the wheel in place without distorting the flange. Always follow the manufacturer's recommendations. Take care to ensure that the spindle speed of the machine will not exceed the maximum operating speed marked on the wheel.

An abrasive wheel may disintegrate or explode during start-up. Allow the tool to come up to operating speed prior to grinding or cutting. The employee should never stand in the plane of rotation of the wheel as it accelerates to full operating speed. Portable grinding tools need to be equipped with safety guards to protect workers not only from the moving wheel surface, but also from flying fragments in case of wheel breakage.

2.3.3 Pneumatic Tools

Pneumatic tools are powered by compressed air and include chippers, drills, hammers, and sanders.

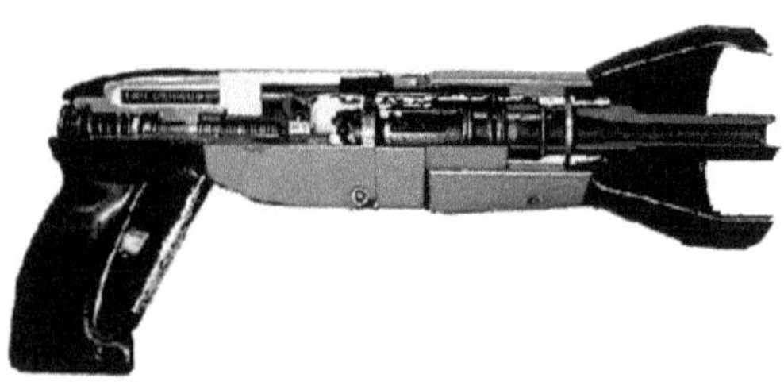

Fig Nail Gun

There are several dangers associated with the use of pneumatic tools.

- First and foremost is the danger of getting hit by one of the tool's attachments or by some kind of fastener the worker is using with the tool.
- Pneumatic tools must be checked to see that the tools are fastened securely to the air hose to prevent them from becoming disconnected. A short wire or positive locking device attaching the air hose to the tool must also be used and will serve as an added safeguard.

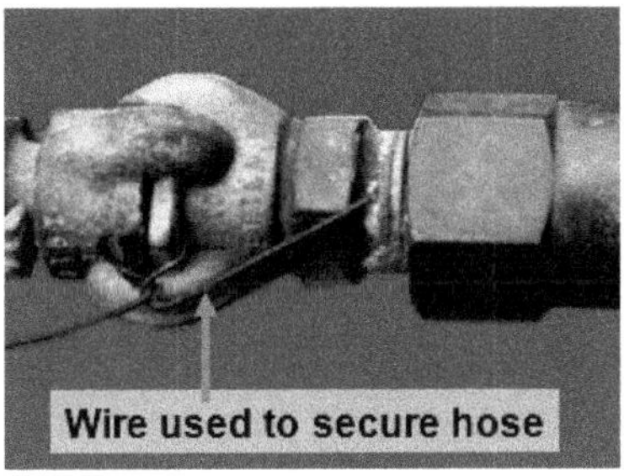

Fig Pneumatic tools with proper fastening

- If an air hose is more than 1/2-inch (12.7 millimeters) in diameter, a safety excess flow valve must be installed at the source of the air supply to reduce pressure in case of hose failure.
- In general, the same precautions should be taken with an air hose that are recommended for electric cords, because the hose is subject to the same kind of damage or accidental striking, and because it also presents tripping hazards.
- When using pneumatic tools, a safety clip or retainer must be installed to prevent attachments such as chisels on a chipping hammer from being ejected during tool operation.

Pneumatic tools that shoot nails, rivets, staples, or similar fasteners and operate at pressures more than 100 pounds per square inch (6,890 kPa), must be equipped with a special device to keep fasteners from being ejected, unless the muzzle is pressed against the work surface.

Airless spray guns that atomize paints and fluids at pressures of 1,000 pounds or more per square inch (6,890 kPa) must be equipped with automatic or visible manual safety devices that will prevent pulling the trigger until the safety device is manually released.

Eye protection is required, and head and face protection is recommended for employees working with pneumatic tools.

- ☆ Screens must also be set up to protect nearby workers from being struck by flying fragments around chippers, riveting guns, staplers, or air drills.
- ☆ Compressed air guns should never be pointed toward anyone. Technicians should never "dead-end" them against themselves or anyone else. A chip guard must be used when compressed air is used for cleaning.
- ☆ Use of heavy jackhammers can cause fatigue and strains. Heavy rubber grips reduce these effects by providing a secure handhold.
- ☆ Workers operating a jackhammer must wear safety glasses and safety shoes that protect them against injury if the jackhammer slips or falls. A face shield also should be used.
- ☆ Noise is another hazard associated with pneumatic tools. Working with noisy tools such as jackhammers requires proper, effective use of appropriate hearing protection.

2.3.4 Liquid Fuel Tools

Fuel-powered tools are usually operated with gasoline. The most serious hazard associated with the use of fuel-powered tools comes from fuel vapors that can burn or explode and also give off dangerous exhaust fumes. The worker must be careful to handle, transport, and store gas or fuel only in approved flammable liquid containers, according to proper procedures for flammable liquids.

Before refilling a fuel-powered tool tank, the user must shut down the engine and allow it to cool to prevent accidental ignition of hazardous vapors. When a fuel-powered tool is used inside a closed area, effective ventilation and/or proper respirators such as atmosphere-supplying respirators must be utilized to avoid breathing carbon monoxide. Fire extinguishers must also be available in the area.

2.3.5 Powder-Actuated Tools

Powder-actuated tools operate like a loaded gun and must be treated with extreme caution. In fact, they are so dangerous that they must be operated only by specially trained employees.

Fig Powder-Actuated Tool

When using powder-actuated tools, an employee must wear suitable ear, eye, and face protection. The user must select a powder level—high or low velocity—that is appropriate for the powder-actuated tool and necessary to do the work without excessive force.

The muzzle end of the tool must have a protective shield or guard centered perpendicular to and concentric with the barrel to confine any fragments or particles that are projected when the tool is fired. A tool containing a high-velocity load must be designed not to fire unless it has this kind of safety device.

To prevent the tool from firing accidentally, two separate motions are required for firing. The first motion is to bring the tool into the firing position, and the second motion is to pull the trigger. The tool must not be able to operate until it is pressed against the work surface with a force of at least 5 pounds (2.2 kg) greater than the total weight of the tool.

If a powder-actuated tool misfires, the user must hold the tool in the operating position for at least 30 seconds before trying to fire it again. If it still will not fire, the user must hold the tool in the operating position for another 30 seconds and then carefully remove the load in accordance with the manufacturer's instructions. This procedure will make the faulty cartridge less likely to explode. The bad cartridge should then be put in water immediately after removal.

If the tool develops a defect during use, it should be tagged and must be taken out of service immediately until it is properly repaired.

Safety precautions that must be followed when using powder actuated tools include the following:

- ☆ Do not use a tool in an explosive or flammable atmosphere.
- ☆ Inspect the tool before using it to determine that it is clean, that all moving parts operate freely, and that the barrel is free from obstructions and has the proper shield, guard, and attachments recommended by the manufacturer.
- ☆ Do not load the tool unless it is to be used immediately.
- ☆ Do not leave a loaded tool unattended, especially where it would be available to unauthorized persons.
- ☆ Keep hands clear of the barrel end.
- ☆ Never point the tool at anyone.

When using powder-actuated tools to apply fasteners, several additional procedures must be followed:

- ☆ Do not fire fasteners into material that would allow the fasteners to pass through to the other side.

- ☆ Do not drive fasteners into very hard or brittle material that might chip or splatter or make the fasteners ricochet.
- ☆ Always use an alignment guide when shooting fasteners into existing holes.
- ☆ When using a high-velocity tool, do not drive fasteners more than 3 inches (7.62 centimeters) from an unsupported edge or corner of material such as brick or concrete.
- ☆ When using a high velocity tool, do not place fasteners in steel any closer than 1/2-inch (1.27 centimeters) from an unsupported corner edge unless a special guard, fixture, or jig is used.

2.3.6 Hydraulic Power Tools

The fluid used in hydraulic power tools must be an approved fire resistant fluid and must retain its operating characteristics at the most extreme temperatures to which it will be exposed. The exception to fire-resistant fluid involves all hydraulic fluids used for the insulated sections of derrick trucks, aerial lifts, and hydraulic tools that are used on or around energized lines. This hydraulic fluid shall be of the insulating type.

The manufacturer's recommended safe operating pressure for hoses, valves, pipes, filters, and other fittings must not be exceeded. All jacks—including lever and ratchet jacks, screw jacks, and hydraulic jacks—must have a stop indicator, and the stop limit must not be exceeded. Also, the manufacturer's load limit must be permanently marked in a prominent place on the jack, and the load limit must not be exceeded.

A jack should never be used to support a lifted load. Once the load has been lifted, it must immediately be blocked up. Put a block under the base of the jack when the foundation is not firm, and place a block between the jack cap and load if the cap might slip.

To set up a jack, make certain of the following:

- ☆ The base of the jack rests on a firm, level surface;
- ☆ The jack is correctly centered;
- ☆ The jack head bears against a level surface; and
- ☆ The lift force is applied evenly.

Proper maintenance of jacks is essential for safety. All jacks must be lubricated regularly. In addition, each jack must be inspected according to the following schedule:

1. for jacks used continuously or intermittently at one site—inspected at least once every 6 months

2. for jacks sent out of the shop for special work—inspected when sent out and inspected when returned
3. for jacks subjected to abnormal loads or shock—inspected before use and immediately thereafter.

2.4 AGRICULTURE WORKSHOP TOOLS

2.4.1. AIR COMPRESSOR

Fig air compressor

A good air compressor will last practically the lifetime of the shop. For durability, avoid oil-less units or aluminium casing compressors. Air compressor size is usually rated first by airflow capacity and pressure is a secondary consideration. As for pressure, you'll need at least 90psi (621kPa) and for heavy-duty work a two-stage compressor with pressures up to 175psi (1,208kPa) is in order.

The most airflow-hungry demands on the compressor come from air tools. For example, a 7-inch (18mm) air-driven angle grinder, an orbital sander or a low pressure spray gun will need up to 8cfm (4l/s). Heavy-duty tools like a one-inch impact wrench or air hammer/chisel will need up to 11cfm (5l/s).

The list of air tools is almost unlimited these days and their air demands are on the packet. Out in the paddock a 25cfm (12l/s) compressor and air lance are essential especially for harvester blow-down.

2.4.2. Workbench

Absolute minimum size would be, say, 4 foot long by 2 foot wide (1.2 X 0.6m).

There's a good case for having a work bench that has both a wood and a metal surface. The wooden section is often easier to work on and precision parts are less likely to be damaged on wood. The bench needs to be solidly mounted. The bench vice should be mounted on the steel top.

2.4.3. Bench Vice

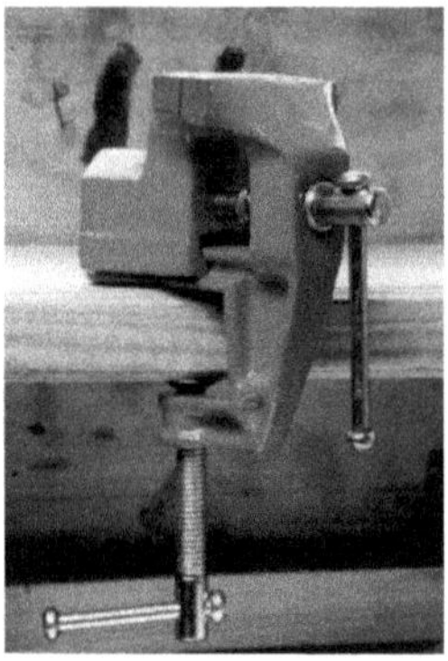

Fig solid bench vice

A minimum jaw size ought to be four inch (100mm). A vice with in-built anvil is mighty handy. In the absence of a vice-integrated anvil, a separate anvil is always valuable. The old blacksmith vices were pretty good – they were anchored directly into floor and could really take a hammering.

In any event the bench vice needs to be securely attached to the bench, preferably located on the bench with space where long steel pieces can be gripped in and on either side of the vice. I have built a couple of useful anvils by welding legs on lengths of inverted railway iron, flat side up.

2.4.4 .Drill Press

Fig Drill press

For precise drilling, a drill press is required not only for precision but also to handle larger hole drilling and to work safer with less strain. It is a second-hand factory machine tools available in any area.

2.4.5. Chop saw

Fig Chop saw

A chop saw does a better job by involving cleaning and fastening than a 'gas wrench' or oxy torch. In this, there's no slag and the cut makes for a better quality and stronger weld.

2.4.6. Angle Grinders

Fig working with angle grinder

At the minimum a 4-inch angle grinder plus a seven-inch angle grinder - these are indispensable.

You use these for shaping things, for weld-prep and for sharpening blades and tools. Bench grinders are useful but can't beat an angle grinder for versatility.

2.4.7. Welder

Nowadays the welding game, old stick welder is a standby, wire welders for slag-free and better penetrating welds. A cheaper face shields as they can give out and get temporarily blinded accidentally by weld flashes if auto-darkening feature fails. An oxy set is versatile but the cost of gas and bottle rental is something to ponder.

2.4.8. Bolt Bins

Fig bolt bin

The good workshop should have wide selection of bolts and nuts. A smart move is to start out with a professional set of bolt containers, as distinct from using old jam tins or odd plastic containers of mixed shapes and sizes.

2.4.9. Hydraulic Press

Here's another place where the bigger the better! A 10-ton press maybe okay, but a broadacre farmer can justify having a 30-ton press.

Having a press can save trips to town to press out bearings, and straighten bent stuff.

2.4.10. Jib Crane/Forklift/FEL

Fig Jib crane

Lifting items is expedited with these tools. The front-end loader (FEL) is good substitute if you don't have space for a jib and winch, or a fork lift in the shop.

Chapter 3

AGRICULTURAL MACHINERIES

3.1 INTRODUCTION

Agricultural machinery is machinery used in farming or other agriculture. There are many types of such equipment, from hand tools and power tools to tractors and the countless kinds of farm implements that they tow or operate. Diverse arrays of equipment are used in both organic and nonorganic farming. Especially since the advent of mechanised agriculture, agricultural machinery is an indispensable part of how the world is fed.

Power for agricultural machinery was originally supplied by ox or other domesticated animals. With the invention of steam power came the portable engine, and later the traction engine, a multipurpose, mobile energy source that was the ground-crawling cousin to the steam locomotive. Agricultural steam engines took over the heavy pulling work of oxen, and were also equipped with a pulley that could power stationary machines via the use of a long belt. The steam-powered machines were low-powered by today's standards but, because of their size and their low gear ratios, they could provide a large drawbar pull.

3.2 TYPES

Combines might have taken the harvesting job away from tractors, but tractors still do the majority of work on a modern farm. They are used to pull implements that till the ground, plant seed, or perform a number of other tasks.

Tillage implements prepare the soil for planting by loosening the soil and killing weeds or competing plants. The best-known is the plow, the ancient implement that was upgraded in 1838 by John Deere. Plows are actually used less frequently today, with offset disks used instead to turn over the soil and chisels used to gain the depth needed to retain moisture.

The most common type of seeder, called a planter, spaces seeds out equally in long rows that are usually two to three feet apart. Some crops are planted by drills, which put out much more seed in rows less than a foot apart, blanketing the field with crops. Transplanters fully or partially automate the task of transplanting seedlings to the field. With the widespread use of plastic mulch, plastic mulch layers, transplanters, and seeders lay down long rows of plastic and plant through them automatically.

After planting, other implements can be used to remove weeds from between rows, or to spread fertilizer and pesticides. Hay balers can be used to tightly package grass or alfalfa into a storable form for the winter months.

Modern irrigation also relies on a great deal of machinery. A variety of engines, pumps and other specialized gear is used to provide water quickly and in high volumes to large areas of land. Similar types of equipment can be used to deliver fertilizers and pesticides.

Besides the tractor, a variety of vehicles have been adapted for use in various aspects of farming, including trucks, airplanes, and helicopters, for everything from transporting crops and making equipment mobile to aerial spraying and livestock herd management.

Cultivators

Agricultural Equipment

Destoner

Agricultural Pipes

Power Sprayer

Power Tiller

Combine Harvester

Agriculture Pumps

Disc Harrow

Paddy Thresher

Cultivating Equipment

Thresher

Potato Planter

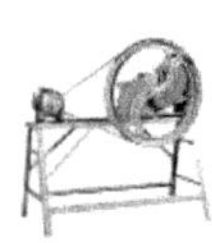

Chaff Cutter

Rotary Tiller

3.3 TOOLS FOR TILLAGE IMPLEMENTATIONS

3.3.1 Plough:

Tillage is the basic operation in farming. It is done to create favourable conditions for seed placement and plant growth. This is done mainly with a plough. A full history of the evolution of plough is not available. Farmers have been using plough since time immemorial. The primitive model might have been a crooked twig or a branch of a tree. The basic components of the plough are a shoe, a share, a body, a handle and a beam.

The shoe and body make one piece in the case of ploughs being used in Kullu, Solan, Shimla, Sirmour, Lahaul & Spiti and Kinnaur district of Himachal Pradesh. The Joint between the shoe and the body is purposely avoided with a view to make the plough more rigid and robust, so that the implement can work on gravely soils with stones and other obstacles which are encountered during the course of ploughing.

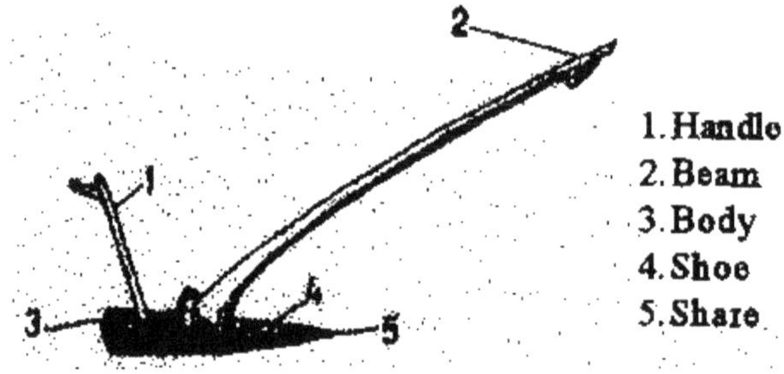

Fig. Traditional plough

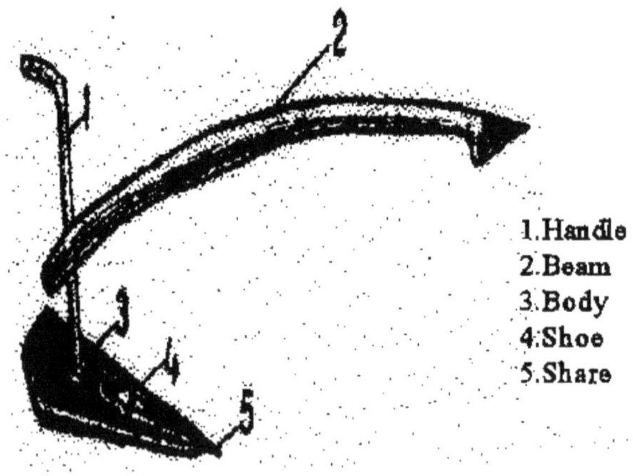

In another type, the shoe and body feature as separate parts. This type is particularly preferred in soils with a greater proportion of sand particles. This type of structure facilitates the replacement of the shoe when it wears out due to the abrasive action of the sand. This type of plough is used in Mandi, Chamba, Una and Kangra districts of Himachal Pradesh. In Sirmour district, the beam of the plough is joined to the handle instead of the body. This is done to provide more clearance under the beam and for easy passage over clods and stubble. The handle is 0.6 to 1 m long, 5 to 7.5 cm thick and 7.5 to' 12.5 cm wide and is fitted to the body of the plough.

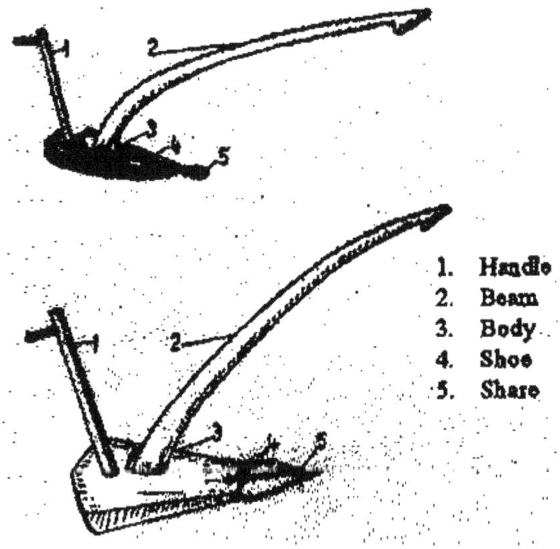

The shoe, used in the plough, can be of different shapes and sizes. Ploughs used in dry lands have shoes which are generally of a triangular section while in wet land cultivation the shoes are of a more flat section.

The share is prepared from a mild steel bar, 0.6 to 7.5 m in length and 1.5 to 2.5 cm in width. The share is fixed to the shoe or body by means of a U-clamp or ring shaped clamp. The share point projects beyond the shoe by 5 to 7.5 cm. Beams generally vary in length from 2.4 to 3 m.

3.3.2 Yoke

The yoke has a projection at the centre to which a beam of implements like plough, leveller and harrow etc. are secured by a rope.

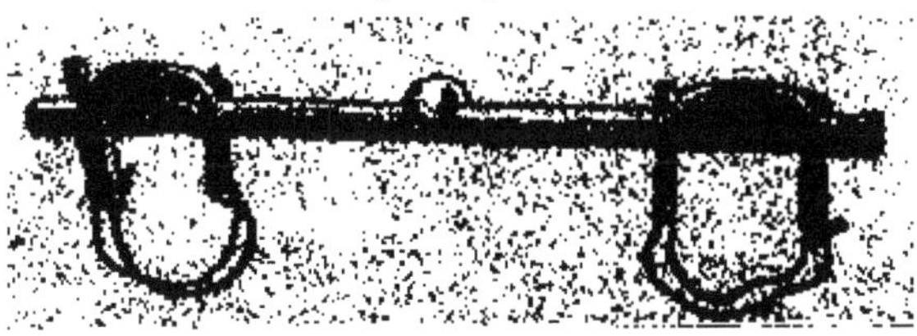

Fig. Neck yoke (jungle)

3.3.3 Leveller

The plank of the leveller is made of any locally available wood and shafts are generally made of bamboo sticks. Extra weight is added to all type of planks by placing stones on it or having person (s) ride on it. As the name suggests, levellers are used for levelling land (Fig. a & b).

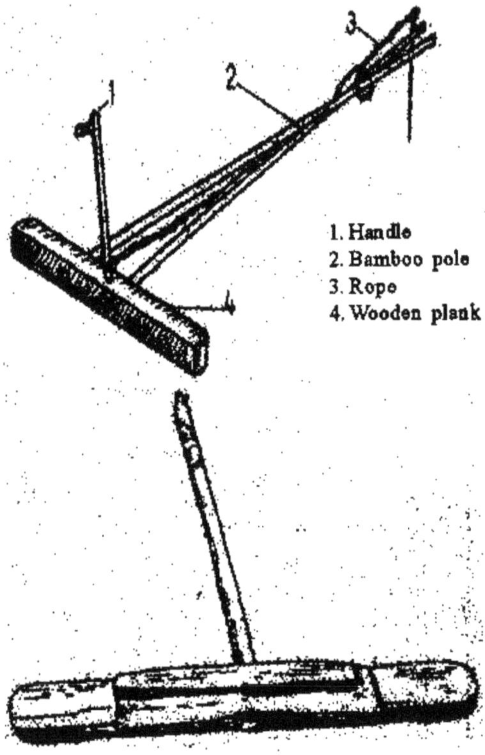

Fig. Leveller (Suhaga)

3.3.4 Harrow

It has a wooden plank to which wood/iron pegs, handle and bamboo shaft are fitted. It is used for breaking soil crust after rain and also for uprooting weeds.

3.3.5 Mallot

It has a wooden block to which a handle is attached. Occasionally, one end of the block is tapering. It is used for the breaking of clods.

3.4 TOOLS FOR INTERCULTURAL OPERATIONS

3.4 1. Khilna

Its handle is made of a branch of sturdy wood and the tool itself is made of iron and is shaped like an arrowhead. It is used for uprooting of weeds.

3.4.2 Kudali

Used for digging and weeding operations, it is made of materials similar to the Khilna with the exception that the iron end is flat.

Harrow (Dandal)

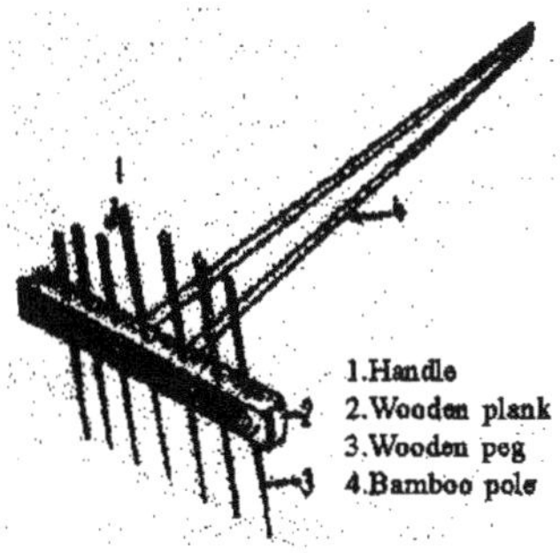

Mallot

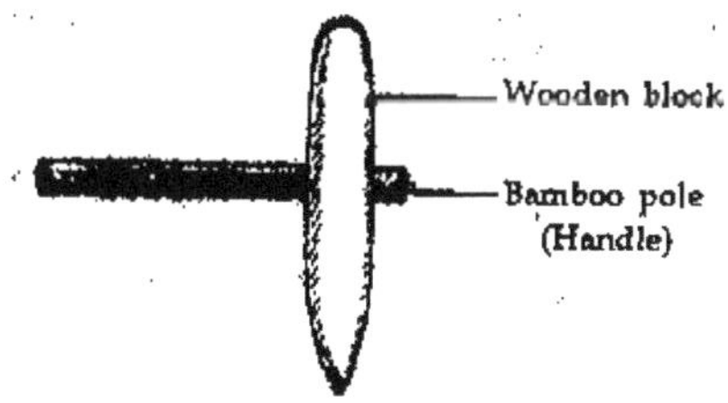

Khilna

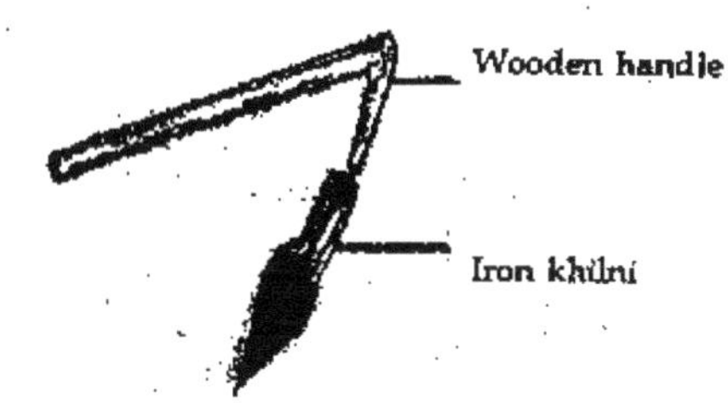

Kudali

Spade (Pharwa)

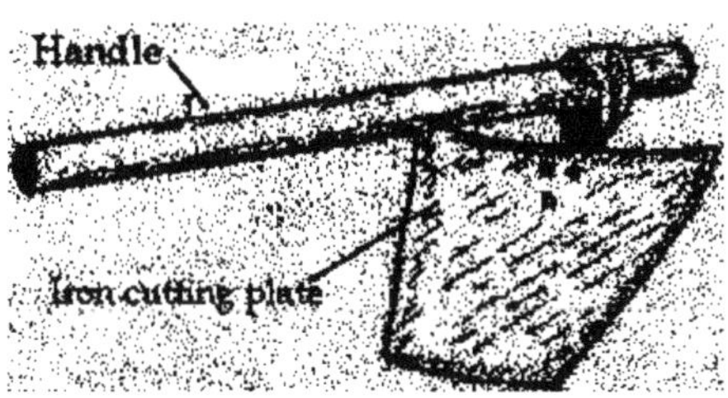

3.5 HARVESTING TOOLS:

The most common type of harvesting implement are small sickle, big sickle, *darat*, *gandasa* and small axe etc., (Fig. 9.9a, b, c & d).

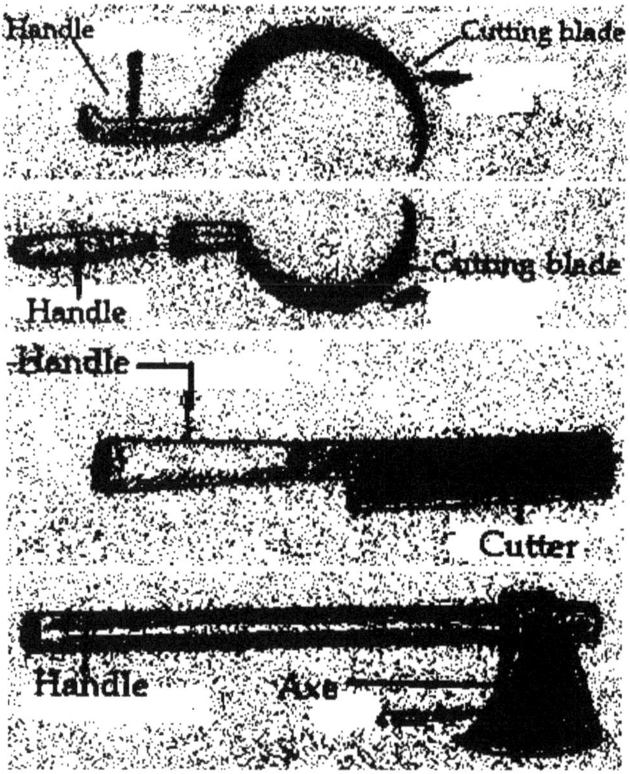

Fig. Harvesting tools

The hand sickle is used to harvest crops like wheat, maize, barley, pulses and grass etc. Big sickle (*Darat*) is used to harvest fodder from trees. *Gandasa* (*chopper*) and axe are used to harvest crops like sugar-cane etc.

3.6 POSTHARVEST TOOLS AND IMPLEMENTATION

3.6.1 Wooden Pin

Is used to remove the outer covering of maize cobs and is fashioned out of a bamboo stick.

3.6.2 Wooden Pole

These are used to detach grains from the maize cobs and grains from other crops through a beating action.

3.6.3 Suhaga

This is used to thresh both wheat and paddy crops according to traditional practice through the rubbing action of the *suhaga* with the tillers of crops.

3.6.4 Bamboo basket (*Kilta*)

This is used for carrying FYM and farm produce.

Fig. Big bamboo basket (Kilta)

3.6.5. Tokri

A small bamboo basket

Fig. Small bamboo basket (Tokri)

3.6. 6. Hand mill

This is used to grind flour and pulses.

Fig. Indigenous hand mill

3.6.7 Winnower

This separates the grain from the husk.

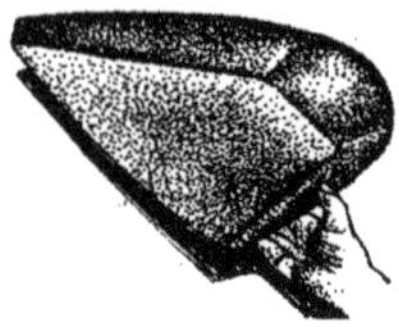

Fig. Winnower (Shoop)

3.6.8. Sieve

This is used for the separation of different types of grains for elimination of alien material.

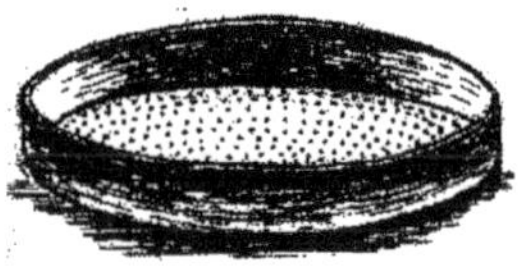

Fig. Sieve (Chanani)

3.6. 9. Sack

It is used to store the farm produce. The traditional sack is made from the skin of sheep and goat.

Fig. Sack

3.6. 10. Shearer

This is used for shearing wool.

3.6.11. Skinner

This is used for taking off the skin of slaughtered goat/sheep.

Fig Shearer (Balwan)

Fig Sheep/goat skinner (Kataru)

3.6.12. Pine

It is a needle collecting tool.

Fig. Tool for collecting pine needles (Kadera)

3.7 TOOLS AND APPROACH

3.7.1 Biological monitoring:

Biological monitoring is one of the three important tools in the prevention of diseases due to toxic agents in the general or occupational environment, the other two being environmental monitoring and health surveillance. The sequence in the possible development of such disease may be schematically represented as follows: source-exposed chemical agent—internal dose—biochemical or cellular effect (reversible) —health effects—disease.

The purpose of biological monitoring of exposure is to assess health risk through the evaluation of internal dose, achieving an estimate of the biologically active body burden of the chemical in question. Its rationale is to ensure that worker exposure does not reach levels capable of eliciting adverse effects. An effect is termed "adverse" if there is an impairment of functional capacity, a decreased ability to compensate for additional stress, a decreased ability to maintain homeostasis (a stable state of equilibrium), or an enhanced susceptibility to other environmental influences.

Depending on the chemical and the analysed biological parameter, the term internal dose may have different meanings. First, it may mean the amount of a chemical recently absorbed, for example, during a single workshift. A determination of the pollutant's concentration in alveolar air or in the blood may be made during the workshift itself, or as late as the next day (samples of blood or alveolar air may be taken up to 16 hours after the end of the exposure period). Second, in the case that the chemical has a long biological half-life—for example, metals in the bloodstream—the internal dose could reflect the amount absorbed over a period of a few months.

Third, the term may also mean the amount of chemical stored. In this case it represents an indicator of accumulation which can provide an estimate of the concentration of the chemical in organs and/or tissues from which, once deposited, it is only slowly released. For example, measurements of DDT or PCB in blood could provide such an estimate.

Finally, an internal dose value may indicate the quantity of the chemical at the site where it exerts its effects, thus providing information about the biologically effective dose. One of the most promising and important uses of this capability, for example, is the determination of adducts formed by toxic chemicals with protein in haemoglobin or with DNA.

Toxic metals and organometallic compounds

Toxic metals and organometallic compounds such as aluminium, antimony, inorganic arsenic, beryllium, cadmium, chromium, cobalt, lead, alkyl lead, metallic mercury and its salts, organic mercury compounds, nickel, selenium and vanadium have all been recognized for some time as posing potential health risks to exposed persons.

Chapter 4

WORKING WITH EQUIPMENTS

4.1 USE OF BASIC HAND SKILLS

Use hand and power tools for nuts and bolts, measuring and marking, cutting and punching, drilling, grinding and welding

4.2 USE AND CARE FOR LIFTING EQUIPMENT.

4.2.1 Introduction

Lifting equipment are work equipment used at work for lifting and lowering loads and includes attachments used for anchoring, fixing or supporting the load which includes equipment such as cranes, vacuum lifting cranes, hoists, scissors lifts, fork lift trucks, passenger lifts, mobile elevating work platforms, vehicle inspection platform hoists, vehicle tail lifts, garage jacks, bath hoists, dumb waiters, pallet trucks, agricultural lifting equipment such as front-end loaders, etc.

Apart from these, lifting equipment also covers chains, ropes, slings, pulleys, eyebolts, shackles, anchor points, karabiners, harnesses and strops; examples of situations include rope and pulley systems on building sites and ropes used for climbing or work positioning in arboriculture.

Equipment that does not have as its principal function a use for lifting or lowering is generally excluded such as three-point linkage on tractors and horizontal conveyors or horizontal winching operations, however, the safety of such equipment would be subject to consideration.

Further it is the responsibility of agriculture mechanic to carry out a suitable and sufficient risk assessment, in order to identify the nature and level of risks associated with lifting operation which are based on following factors:

- ☆ type of load being lifted, its weight, shape and what it consists of;
- ☆ risk of a load falling or striking a person or other object and the subsequent consequences;
- ☆ risk of lifting equipment failing or falling over while in use and subsequent consequences.

4.2.2 Suitability of Lifting Equipment

By selecting the correct lifting equipment for particular tasks and processes it is possible to eliminate or reduce many risks to the health and safety of people in the workplace. Issues to be considered when assessing suitability include:

- ☆ ergonomic design and its compatibility with users;
- ☆ material of manufacture and their suitability for conditions of use;
- ☆ equipments initially integrity
- ☆ place in which it will be used;
- ☆ means of access/egress need to be safe and suitable.
- ☆ need to minimise risks from slips, trips and falls from any part of the lifting equipment (e.g. cover or fence any floor openings, suitable edge protection where there is a risk of falling more than 2m, or less then 2m where there are other relevant risk factors involved.
- ☆ adequate protection for operators from adverse environmental conditions (e.g. high winds).

It is noted that lifting equipments used should be sufficiently strong, stable and suitable for the proposed use, with an appropriate factor of safety against failure built in. Similarly the load and anything attached to it must be suitable and of adequate strength. Where appropriate, sufficient resistance to overturning must be provided to be put into place and operating effectively before the equipment is used (e.g. outriggers, stabilisers, use of ballast etc.). Where mobile lifting equipment is fitted with pneumatic tyres, they must be inflated to the correct pressure before lifting operations commence. Agriculture technician must provide adequate means to check regarding the significant risk involved in overloading and/or overturning which could arise from use of equipment which covers device rated capacity indicators and limiters.

4.2.3 Positioning and Installation

Lifting equipment must be positioned or installed to as low as reasonably practicable prevent the risk of the equipment or the load striking people, or the risk of the load drifting, falling freely or being released unintentionally. It is therefore essential that lifts over people are minimised and unauthorised access to the work area prevented. Equipment should also be installed and positioned to prevent crushing when the equipment is in its extreme position. Wherever possible, hooks and other similar devices used for lifting should have either safety catches fitted or be shaped to prevent the accidental displacement of the sling.

Where two or more items of lifting equipment are used they, or their loads, should be prevented from coming into contact with each other. Agriculture technician must ensure suitable devices (e.g. gates) are provided to prevent persons falling down a lift shaft or hoist-way. Such gates should be fitted with efficient interlocking systems which prevent lift movement until the gates are closed.

4.2.4 Marking of Lifting Equipment

Lifting machinery and associated accessories should be marked with their safe working load (SWL). Where it is not practicable to mark the equipment itself (e.g. ropes, slings, karabiners, strops, and harnesses for rope access), a coding system should be used to provide the user with the SWL (e.g. colour coding, label). This also applies where the SWL is dependent upon varying configurations (e.g. fork lift truck fitted with attachments). Where a significant hazard arises from the use of lifting equipment, appropriate equipment or devices such as rated capacity indicators/ limiters should be provided.

4.2.5 Organisation of Lifting Operations

The agriculture technician should ensure that every lifting operation (including lowering loads) should be:

a) properly planned by a competent person i.e. person has adequate practical and theoretical knowledge and experience of lifting operations. The plan must address identified risks, resources required, procedures, responsibilities etc);

b) appropriately supervised (proportionate to the risk - taking into account the personnel involved);

c) carried out in a safe manner.

Plans should consider issues such as:

- ☆ weight, shape, centre of gravity and lifting points of the load;
- ☆ where the load is and where it is to be moved to (including route, visibility etc);

- ☆ how often the lifting equipment will be used;
- ☆ how loads will be attached/detached, compatibility of lifting accessories, safe working loads of equipment, overloading;
- ☆ the environment in which the lifting is to take place;
- ☆ likelihood of overturning, proximity hazards;
- ☆ number of persons available, their knowledge, training and experience, requirement for supervision;
- ☆ pre-use checks;
- ☆ storage of lifting equipment to prevent damage or deterioration;
- ☆ if lifting persons, suitability of equipment, means of communication, failure of equipment, reliable means of rescue etc.

4.2.6 Examination and Inspection

It is the duty of agriculture mechanic to ensure that lifting equipment is thoroughly examined and inspected. The risks arising from failure will determine how thorough the examination needs to be. Thorough examination may be needed at several points in the life of lifting equipment.

Before using lifting equipment, a thorough examination must be carried out unless there is physical evidence available to show that it is safe to use which could include if the equipment:

- ☆ has not been used before and there is documentation drawn up within the 12 months prior to its first use, an initial test certificate in the form of an "EC Declaration of Conformity" or
- ☆ has been obtained from another organisation (e.g. hired or borrowed) and is accompanied by evidence that the necessary thorough examination has been carried out, (e.g. a copy of the latest test certificate is available).

Such evidence or competent person's examination report must be kept until the faculty/service/department ceases to use the lifting equipment, or in the case of accessories, 2 years.

Additionally a thorough examination is required after substantial or significant modification or repair. This covers installation in a new location or a reconfiguration.

4.2.7 Periodic examinations once equipment is in service

- ☆ Lifting equipment must also be thoroughly examined at regular intervals for smoothness of work. Lifting equipment in use must be thoroughly examined at least every 12 months;

- Accessories (chains, slings, ropes, etc.) must be thoroughly examined at least every 6 months;
- Equipment used for lifting people must be thoroughly examined at least every 6 months;
- For the above, the examination is in accordance with a written examination scheme and is to identify any remedial action in good time before deterioration affects safety.

4.2.8 Reports and Defects

It is the duty of agriculture mechanic to:

- notify any dangerous defects;
- report in writing made as soon as is practicable
- to concern
- to any person from whom the equipment has been hired or leased;
- where the defect involves an existing or imminent risk of serious personal injury;
- ensure that lifting equipment is not used before the defect is remedied, or after a time specified in a report and before the defect is remedied.

4.3 KNOWLEDGE ABOUT CLUTCH

4.3.1 Introduction

A clutch is a mechanical device which engages and disengages power transmission especially from driving shaft to driven shaft. In the simplest application, clutches connect and disconnect two rotating shafts (drive shafts or line shafts). In these devices, one shaft is typically attached to an engine or other power unit (the driving member) while the other shaft (the driven member) provides output power for work. While typically the motions involved are rotary, linear clutches are also possible.

In a torque-controlled drill, for instance, one shaft is driven by a motor and the other drives a drill chuck. The clutch connects the two shafts so they may be locked together and spin at the same speed (engaged), locked together but spinning at different speeds (slipping), or unlocked and spinning at different speeds (disengaged).

Various materials have been used for the disc-friction facings, including asbestos in the past. Modern clutches typically use a compound organic resin with copper wire facing or a ceramic material. Ceramic materials are typically used in heavy applications such as racing or heavy-duty hauling, though the harder ceramic materials increase flywheel and pressure plate wear.

In the case of "wet" clutches, composite paper materials are very common. Since these "wet" clutches typically use an oil bath or flow-through cooling method for keeping the disc pack lubricated and cooled, very little wear is seen when using composite paper materials.

Friction-disc clutches generally are classified as push type or pull type depending on the location of the pressure plate fulcrum points. In a pull-type clutch, the action of pressing the pedal pulls the release bearing, pulling on the diaphragm spring and disengaging the vehicle drive. The opposite is true with a push type, the release bearing is pushed into the clutch disengaging the vehicle drive. In this instance, the release bearing can be known as a thrust bearing.

A clutch damper is a device that softens the response of the clutch engagement/ disengagement. In automotive applications, this is often provided by a mechanism in the clutch disc centres. In addition to the damped disc centres, which reduce driveline vibration, pre-dampers may be used to reduce gear rattle at idle by changing the natural frequency of the disc. These weaker springs are compressed solely by the radial vibrations of an idling engine. They are fully compressed and no longer in use once the main damper springs take up drive.

4.3.2 Clutch Assembly

A clutch assembly is a device which, when disengaged, interrupts the power from an engine or motor to a gearbox. This allows the machine to engage or disengage gears, start or stop, and move off smoothly.

Fig Inspecting clutch assembly

A simple clutch assembly has three main parts; the clutch disk, the pressure plate and the release bearing. The pressure plate clamps the clutch disk up against the engine's flywheel, thereby transmitting the power from the engine through the clutch to the gearbox. To interrupt this flow, the operator presses the clutch pedal and releases the pressure plate, meaning the clutch disk is no longer clamped against the flywheel and the drive is interrupted.

There are generally two common faults with clutches. The first is clutch slippage, recognised as a sudden increases in engine revs without the proportional increase in ground speed, especially while pulling a heavy load. In this case, either the pressure plate is getting weak, or the friction surfaces of the clutch disk are becoming worn.

The second fault is a difficulty in selecting a gear, often without clutch slippage. In this case, the release bearing is generally at fault. What happens is the release bearing becomes worn over time, and is no longer able to push on the pressure plate to release the clutch and interrupt the power from the engine.

4.4 IDEA ABOUT DRIVE SHAFT

4.4.1 Introduction

A drive shaft, driveshaft, driving shaft, propeller shaft (prop shaft), or Cardan shaft is a mechanical component for transmitting torque and rotation, usually used to connect other components of a drive train that cannot be connected directly because of distance or the need to allow for relative movement between them.

As torque carriers, drive shafts are subject to torsion and shear stress, equivalent to the difference between the input torque and the load. They must therefore be strong enough to bear the stress, while avoiding too much additional weight as that would in turn increase their inertia.

To allow for variations in the alignment and distance between the driving and driven components, drive shafts frequently incorporate one or more universal joints, jaw couplings, or rag joints, and sometimes a splined joint or prismatic joint.

4.4.2 Signs of bad driveshaft

- ☆ Vibrations from under the vehicle: A common symptom of a failing driveshaft is an intense shaking coming from underneath the vehicle. Worn out u-joints or bushings can cause the driveshaft to vibrate. If you don't get the u-joints or bushings serviced, it can lead to further damage to other drive train components. Vibrations caused by tire balance issues are speed sensitive while driveshaft vibrations aren't.
- ☆ Difficulty turning: If you're having trouble making turns, it could be a driveshaft issue. A failing driveshaft can prevent the wheels from properly turning, making it difficult to control the vehicle.

- ☆ Loud clunking noise: If you hear a loud clunking noise when shifting your vehicle, this is another sign of driveshaft damage. The likely culprit is a worn u-joint.
- ☆ Shuddering upon acceleration: If you experience shuddering when accelerating from a stop or low speed, a loose u-joint or bad center bearing could be to blame.
- ☆ Squeaking noise: Not just annoying, a squeaking noise at low speeds can be caused by a u-joint that needs lubricating. A little bit of grease will get rid of the noise.
- ☆ Clicking or knocking noise: Can indicate a worn or failing CV joint.

4.5 REMOVING UNIVERSAL JOINTS

A universal joint (universal coupling, U-joint, Spicer or Hardy Spicer joint, or Hooke's joint) is a joint or coupling connecting rigid rods whose axes are inclined to each other, and is commonly used in shafts that transmit rotary motion. It consists of a pair of hinges located close together, oriented at 90° to each other, connected by a cross shaft. The universal joint is not a constant-velocity joint.

A clicking or clunking sound from under your car may indicate a worn universal joint at the drive shaft. Universal joint replacement may seem like a tall task, but it can be done by a novice mechanic with regular tools.

1. Identify the Drive Shaft

Fig Identifying Drive shaft

The first step in replacing your u-joint is getting at it. To do this, you'll need to remove the drive shaft. On some vehicles, you remove the drive shaft by removing the universal joint. If your vehicle is like this, follow the instructions that follow and your drive shaft will drop.

If your drive shaft attaches with axle type bolts (a circle of bolts with either hex or Allen heads), proceed to remove that end of the driveshaft like the photo above. If your driveshaft does not have this type of end, move to the next step to see how to remove it.

2 Remove the Drive Shaft

Fig Disconnecting drive shaft by removing the holding bolts

The other end of your drive shaft may look like this. If it does, you can remove it by removing the two bolts the hold the two halves of the u-joint together. The drive shaft will drop easily once this is removed.

3 Remove Your Universal Joints Snap Rings or C-Clips

Fig Removing the snap rings on a standard universal joint with pliers

There are two types of clips that hold a universal joint together. Spicer Snap Rings are one type and are pictured as shown. The other type is standard C-clips and they are quite easy to remove as well. The steps are the same whichever clip your vehicle uses.

To remove the snap rings, squeeze the ends together with pliers or a special removal tool. They should come out easily. If they have become corroded, you may need to soak the u-joint with some penetrant.

4 Press the Bearing and Joint Out

Fig Tapping the bearing race through

With the drive shaft out of the car or truck, the next step is to remove the joint and its bearings. You'll need to mount the end in a vice, then tap the bearing through far enough to get the joint to drop out. In most cases, a spark plug socket is the perfect tapping tool. Firmly tap the top of your socket with a hammer until the whole thing is pressed far enough in to be able to pop it out.

5 Remove the Bearing Cup from Universal Joint

Fig Bearing cup removal, u-joint repair.

With the joint tapped through, remove the bearing cup (it looks like a cup and is full of grease). This will enable you to remove the loose joint section fully.

6 Insert New UV Joint

Fig Reinstallation of universal joint parts

Get that old joint center out of there and put the new piece in. There's a chance that your new center will assemble a little differently than your old one came out. This is ok. Most replacement joints look a little different. Before you reinstall, lubricate every part you can reach with grease. Don't overpack the assembly with grease as this can cause parts failure, but apply a thin coat to everything. Now's the time to reinstall the bearing cups if your new part comes apart in this manner.

7 Install the New Clips

Fig Installing the new snap rings or c-clips

Now that your assembly is almost back together, you can reinstall the snap rings or c-clips your universal joint uses. The rebuild kit came with new clips, and they look great, don't they?

8 Reinstall Your Drive Shaft

To use the old auto repair adage, installation is the reverse of removal. You're ready to reinstall your drive shaft with the new universal joint ready to provide years of service.

4.6 MACHINERY SELECTION

4.6.1 Introduction

Equipment that works best one year may not work well the next because of changes in weather conditions or crop production practices. Improvements in design may make older equipment obsolete. And, the number of acres being farmed or the amount of labor available may change.

Because many of these variables are unpredictable, the goal of the good machinery manager should be to have a system that is flexible enough to adapt to a range of weather and crop conditions while minimizing long-run costs and production risks.

4.6.2 Selection criteria

First, each piece of machinery must perform reliably under a variety of field conditions or it is a poor investment regardless of its cost.

Tillage implements should prepare a satisfactory seedbed while conserving moisture, destroying early weed growth and minimizing erosion potential. Planters and seeders should provide consistent seed placement and population as well as properly apply pesticides and fertilizers. Harvesting equipment must harvest clean, undamaged grain while minimizing field losses.

The performance of a machine often depends on the skill of the operator, or on weather and soil conditions. Nevertheless, differences among machines can be evaluated through field trials, research reports and personal experience.

Machinery Costs

Once a particular type of tillage, planting, weed control, or harvesting machine has been selected, the question of how to minimize machinery costs must be answered. Machinery that is too large for a particular farming situation will cause machinery ownership costs to be unnecessarily high over the long run; machinery that is too small may result in lower crop yields or reduced quality.

Ownership Costs

Machinery ownership costs include charges for depreciation, interest on investment, property taxes, insurance and machinery housing. These costs increase in direct proportion to machinery investment and size.

Operating Costs

Operating costs include fuel, lubricants and repairs. Operating costs per acre change very little as machinery size is increased or decreased. Using larger machinery consumes more fuel and lubricants per hour, but this is essentially offset by the fact that more acres are covered per hour. Much the same is true of repair costs. Thus, operating costs are of minor importance when deciding what size machinery is best suited to a certain farming operation.

Labor Cost

As machinery capacity increases, the number of hours required to complete field operations over a given area naturally declines. Estimates of time requirements for completing machinery operations are found in AgDM Information File A3-24 Estimating the Field Capacity of Farm Machines.

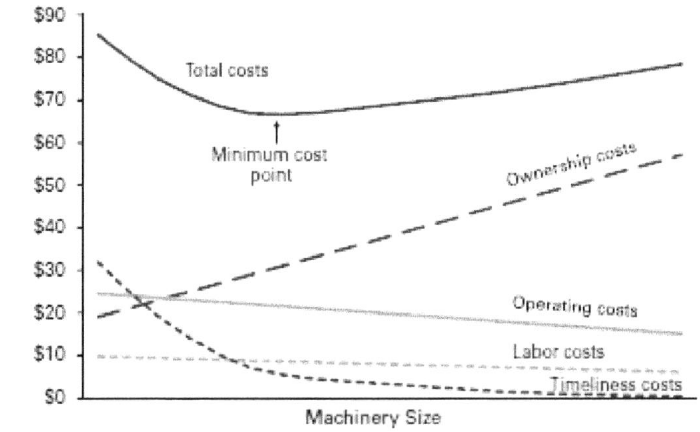

Figure. Effect of increasing machinery size on machinery costs

If hourly or part-time hired labor operates machinery, it is appropriate to use the wage rate paid, plus the cost of any other benefits which may be provided, as the labor cost. If the farmer-owner or a hired worker who is paid a fixed wage operates machinery, then it is proper to value labor at its opportunity cost, or the estimated return it could earn if it were used elsewhere in the farm business, such as in livestock enterprises.

Timeliness Costs

In many cases, crop yields and quality are affected by the dates of planting and harvesting. This represents a "hidden" cost associated with farm machinery, but an important one nevertheless. The value of these yield losses is commonly referred to as "timeliness costs."

Total Machinery Costs

Fig above illustrates the effect that changes in machinery size have on each type of cost in a typical situation. For very small machinery (relative to crop acres), a slight increase in machinery size can lower timeliness and labor costs significantly, enough to more than offset the higher fixed costs. However, as machinery size continues to increase, the timeliness cost savings diminish, and eventually total costs begin to rise. One objective of machinery selection, then, is to select machinery in the size range where total machinery costs are lowest.

4.6.3 Factors Affecting Size of Machinery required

Machinery recommendations must be based on the characteristics of each individual farm. The following factors influence machinery selection, and are discussed in order of importance.

Number of Crop Acres

As more crop acres are farmed, larger-scale machinery is needed to ensure that planting and harvesting are completed in a timely fashion. An alternative is to acquire a second unit of some machines, if an additional tractor and operator are available.

Labor Supply

The number of acres that can be completed each day is the most critical measure of machinery capacity, more than machine width or acres completed per hour. Increasing the labor supply by hiring extra operators or by working longer hours during critical periods may be a relatively inexpensive way of stretching machinery capacity. In addition, the cost of additional labor only needs to be incurred in those years in which it is actually used, while the cost of investing in larger machinery becomes "locked in" as soon as the investment is made. On the other hand, extra labor may not always be available when needed, and working long hours over several days can present a safety hazard.

Tillage Practices

The number of field days needed before planting is completed depends partly on the number of separate operations completed on each acre. Reducing the number of tillage practices performed or performing more than one practice in the same trip effectively decreases the amount of machinery capacity needed to complete field operations on time. Of course, machinery cost savings from reduced tillage must be compared to possible increased chemical costs and effects on yields.

Crop Mix

Diversification of crops tends to spread out the periods when timely completion of field operations is critical. For example, yield reductions due to late planting begin later for soybeans than for corn. Harvesting can also be completed over a longer time period. Thus, growing more than one or two crops reduces the machinery capacity needed for a given number of crop acres. However, it may also require purchasing additional types of machinery, especially for harvesting.

Weather

Weather patterns determine the number of days suitable for fieldwork in a given time period each year. Although actual weather conditions cannot be predicted far enough in advance to be used as an aid to machinery selection, past weather records can be used as a guide. AgDM Information File A3-25 Fieldwork Days in Iowa (PM 1874) lists the number of suitable field days expected for different periods of the year in each of the nine crop reporting districts in Iowa. As a rule of thumb, weather is suitable for field work about 60 percent of the time in the spring and about 75 percent of the time in the fall. This does not take into account time off for holidays, Sundays or other occasions. Machinery selection should be based on long-run weather patterns even though it results in excess machinery capacity in some years and insufficient capacity in other years.

Risk Management

Fluctuations in the number and occurrence of suitable field days from year to year cause timeliness costs to vary even when the machinery set, number of crop acres and labor supply do not change. Investing in larger machinery can reduce the variability of net machinery costs by ensuring that crops are planted and harvested on time even in years in which there are few good working days. Machinery fixed costs would be higher with larger machinery, but they would not fluctuate as long as the machinery set did not change. Farmers with high fixed cash flow needs, such as land mortgage payments, may be willing to pay more (in higher fixed machinery costs) than other operators for the "insurance" of not suffering substantial yield losses due to late planting and harvesting in certain years

4.6.4 Size of Machinery

One way to measure the capacity of a set of machinery is by the number of work days required to complete field operations. This depends on the number of crop acres, machinery operations performed, size of the machinery in use, and availability of labor.

In research conducted recently at ISU, total machinery costs for Iowa grain farms, including the value of timeliness losses, were estimated for a number of different machinery combinations. The effects of variations in the number of crop acres and labor supply, were compared.

Under each set of circumstances the machinery set for which total costs were lowest, including yield losses due to poor timeliness, was identified. In some cases, several machinery sets gave nearly identical minimum costs. In approximately 80 percent of the cases tested, the least-cost machinery sets were able to complete all tillage and planting operations in about 20 to 25 field days. A good rule of thumb for farmers who wish to have sufficient machinery capacity to reduce risk, as well as maintain total costs at a low level, is to be able to complete tillage and planting in about 20 field days.

Where less than one full-time person is available to operate machinery, a goal of 25 to 30 days for completing planting and tillage will most often minimize costs. On the other hand, farms with 2 or 3 full-time machinery operators available could aim to complete this work in less than 20 days.

The machinery sets which minimized total machinery costs were most often able to complete harvesting of corn and soybeans in 25 to 30 field days. As with spring work, operators for whom risk reduction is important should use the lower end of this range as a goal, although yield losses from late harvesting are generally not as severe as from late planting.

A number of different machinery combinations may allow fieldwork to be completed in the same number of days. In putting together a machinery set, it is also important to correctly match machinery sizes and tractor power. Using tractors with horsepower in excess of that required for the implement being pulled results in excessive depreciation and interest costs, while using too little horsepower may cause faster engine wearout.

Chapter 5

CHECKING OF AGRICULTURAL EQUIPMENT

5.1 AGRICULTURAL MACHINERY

Agriculture machineries are designed to till the soil and render it more suitable for crop growth, to sow seeds, to apply agricultural chemicals for improved plant growth and control of pests and diseases, and to harvest and store the mature crops. There is an extremely wide variety of agricultural machines, but all are essentially a combination of gears, shafts, chains, belts, knives, shakers and so on, assembled to perform a certain task. These parts are usually suspended in a frame which may be either stationary or, as is more often the case, mobile and designed to perform the desired operation while moving across a field.

5.2 GROUPS OF AGRICULTURAL MACHINERY

The major groups of agricultural machines are: soil tillage machines; planting machines; cultivating machines; forage harvesting machines; grain, fibre, vegetable, and fruit and nut harvesting machines; agricultural chemical applicators; transport and elevating machines; and sorting and packaging machines.

5.2.1 Soil tillage machines

These include ploughs, tillers, subsoilers, harrows, rollers, levellers, graders and so on. They are designed to turn, agitate, level and compact the soil to prepare it for planting. They may be small in size and require only a small power source (as in the case of a one-person roto-tiller for tilling a rice paddy), or they may be large and require a considerable power source (as in the case of a combined subsoiler, drill and harrow).

5.2.2 Planting machines

These include planters, drills, broadcast seeders and so on and are designed to take seeds from a hopper or bin and insert them in the soil at a predetermined depth and spacing or spread them uniformly over the ground. Planters may be of simple design and comprise a single-row seeding mechanism, or they may be highly complex (as is the case with the multi-row planter with attachments that simultaneously add fertilizer, pesticides and herbicides).

5.2.3 Cultivating machines

These include rotary hoes, cultivators, weeders (mechanical and flame) and so on. They are used to eradicate undesirable weeds or grasses which compete with the plant for soil moisture and make the harvest of the crop more difficult. They also improve the soil tillage so as to make it more absorptive of rain.

5.2.4 Forage harvesting machines

These include mowers, choppers, balers and so on and are designed to sever the stems of roughage crops from their roots and prepare them for storage or immediate use. The machines also vary in their complexity: the simple mower merely cuts the crop, whereas the chopper will not only separate the stalk from the root but will also chop the entire plant into small pieces and load it into a vehicle, which may be a towed wagon. Crimpers, which crush or break the stems of plants, are often used to expedite the field-drying process of fodder crops to prevent spoilage, especially of legumes that will be placed in dry storage or baled. Pelleting machines are used to compress fodder crops into compact cubes for mechanical feeding of livestock. Balers are used to compress fodder into square or round bales to facilitate storage and handling. Some bales are small enough (20 to 40 kg) to handle manually, while others may be so large (400 to 500 kg) as to require mechanical handling systems.

5.2.5 Grain and fibre harvesting machines

These include reapers, binders, corn pickers, combines, threshers and so on. They are used to remove the ripe grain or fibre from the plant and place it in a bin or bag for transport to the storage area. Grain harvesting may involve the use of a number of machines, such as a reaper or binder to cut the standing grain, a wagon or truck to transport the crop to the threshing or separating machines and vehicles to transport the grain to a storage area. In other cases many of these functions may be performed by a single machine, the combine harvester, which cuts the standing grain, separates it from the stalk, cleans it and collects it in a bin, all while moving through the field. Such machines will also load the grain into transport vehicles. Some machines such as cotton pickers and corn pickers may operate selectively and remove only the grain or fibre boll from the stem or stalk.

Figure Combine for harvesting wheat without an enclosed cabin

5.2.6 Vegetable harvesting machines

These include diggers and lifters, and are designed either to dig the crops from the earth and separate them from the soil or to lift or pull the plant free. The potato digger, for example, may form part of a potato combine comprising a sorting, grading device, polisher, bagger and elevator. At the other extreme is the simple two-wheeled, bladed sugar-beet lifter which is followed by hand labourers.

5.2.7 Fruit and nut harvesting machines

These machines are used to harvest berries, fruit and nuts. They may be as simple as a tractor-mounted, vibrating tree shaker which separates the ripe fruit from the tree. Or they may be as complex as the ones which harvest the fruit, catch the falling fruit, place it in a storage container and later transfer it to transport vehicles.

5.2.8 Transport and elevating machines

These also vary considerably in size and complexity ranging, for example, from a simple wagon comprising merely a platform on wheels to a self-loading and stacking transport unit. Inclined chain, flight or belt conveyors or other mechanical handling devices are used to move bulky material (hay, straw, ear corn and so on) from wagon to storage or from one location in a building to another. Screw conveyors are used to move granular material and grain from one level to another, and blowers or pneumatic conveyors are used to move light materials horizontally or vertically.

5.2.9 Agricultural chemical applicators

These are used to apply fertilizers to stimulate plant growth or herbicides and pesticides to control weeds and pests. The chemicals may be liquid, powdered or granular, and the applicator distributes them either by pressure through a nozzle or by centrifugal force. Applicators may be portable or vehicle-mounted; the use of aircraft for chemical application is growing rapidly.

5.2.10 Sorting and packaging machines

These machines are usually stationary. They may be as simple as a fanning mill, which grades and cleans grain merely by passing it over a series of screens, or as complex as a seed mill, which will not only grade and clean but also, for example, separate different types of seeds. Packaging machines usually form part of a sophisticated grading system. They are used primarily for fruit and vegetables and may wrap the produce in paper, bag it or insert it into a plastic container.

5.2.11 Power plants

Electric motors may be used to drive stationary equipment permanently located near a mains supply; however, since many agricultural machines are mobile and must operate in remote areas, they are usually powered by an integral petrol engine or by a separate engine such as that of a tractor. Power from a tractor may be transmitted to the machine via belt, chain, gear or shaft drives; most tractors are fitted with a power take-off coupling specially designed for this purpose.

5.3 TESTING OF MACHINERY

5.3.1 Introduction

The most difficult and vexing problem that the Agricultural Machinery Administration is faced with in farm machinery testing is the development of a suitable test technique for the lair and adequate evaluation of the performance characteristics of a farm machine. The testing of farm machinery on a scientific basis is a comparatively recent science and because of this, there exists very little in terms of "ready made" testing instruments. It is therefore necessary to design new instruments of measurement or adapt instrumentation designed primarily for the aircraft or automotive industries.

It is rather distracting to note that up until very recent years, the designing of farm machinery with the exception of tractors, has been carried out by manufacturers on a very general basis. The primary objective of design engineers has been to improve an acceptable product, while placing a comparatively small amount of emphasis on the investigation of new deigns.

The technique employed is comparatively straight forward, involving the performance comparison of a newly designed component with a component that has already proven acceptable in use by the purchaser. New designs are regarded as superior only if they perform in the field in a manner superior to the component they are replacing.

Manufacturers in the past year or more have come to depend more on the specific measurement of the field loadings of various component parts and are basing their design calculations on this data. Time is always pressing in the testing of farm machinery, and personnel are constantly faced with the problem of where to strike the balance between prolonged and detailed testing to secure adequacy and full information on one hand, and on the other, perfunctory observation for the sake of a quick opinion.

It must also be noted that unless the test conditions can be defined in some repeatable way, it is difficult to assess the merits of a particular machine or to make comparisons between the performance of any two machines tested on different dates or in different seasons. Because of this difficulty, it is necessary to introduce such devices in a technique as "comparison" or "Standard machines", or "Series testing." These two techniques developed originally to represent difficult problems in themselves. However, they can usually be relied upon to supplement the brief descriptions, photographs and whatever scientific measurements are feasible to present an unbiased and factual report.

The comparison or standard machine technique involves the use of some particular machine that is fairly well accepted on the market in the area, as a "baseline" unit against which one may compare the function al performance of the machine under test, thus minimizing the problem of variation in conditions year after year.

The series testing of machines involves the concurrent testing of a number of different makes of machines of the same general type. This technique represents more field problems since a number of machines are being handled at one time as well as the various allied equipment that is required. It is also more costly in as much as a good deal of the instrumentation must be duplicated on each machine.

These problems are compensated however, by the ability of the observer to evaluate the performance characteristics of a number of machines under identical field conditions at the same time and under the same field loadings.

Although it is necessary to develop a test technique for each type of machine that is to be tested, there are a number of general factors that should be included in all techniques which may be generally stated as follows:

1. An assessment of the quality of the work done by the machine.
2. An assessment of the quantity or rate of doing work which will usually involve some measurement of the machine's capacity interpretable by the user.
3. A measurement of the power requirements in terms of horse power, including that required to power the integral parts of the machine as well as the traction drives if applicable. This very of ten includes a measurement of the fuel consumption. This should also include the P.T.O. horsepower requirements.
4. An assessment of or observations with regard to, the labour requirements of a machine with some measurement of the effort required to operate the various work centres of more than one.
5. An evaluation either measured or observed, of the handling characteristics or ease of operation of the machine including the effort required to operate mechanisms such as the clutch, brake and central levers. This usually includes an assessment of the normal time required for service and adjustment.
6. A recording of the structural history during the time the machine was on test. This will include breakage, distortion and fatigue failures, and these will always be associated with hours of use.
7. An evaluation of the safety features of the machine including the protective devices for the machine itself as well as the operator. This will also include a determination of the center of gravity of each machine to determine its safety on slopes of 20%.

5.3.2 Equipment performance checking

Test and measurement equipment play a critical role in most production and development processes. These instruments are used to make decisions regarding the effectiveness of processes and to ensure product quality. It is no surprise, therefore, that the need for regular calibration of test and measurement equipment is included as part of most major quality standards and are a commonly reviewed area for auditors, whether they are from an internal department, accreditation body, or a regulator such as the FDA. Most quality systems will also define the actions that must be followed in the case of an instrument being found out of specification during a calibration. What sometimes does not get enough attention is the selection of test equipment to reduce the risk of an out-of-tolerance condition affecting the process.

There is a fine balance to be considered when selecting equipment. Cost is always an important consideration, but there are other critical issues also like appropriateness of the instrument to the process. It is critical that the instrument selected can meet the measurement requirements of the process not only on the day it is delivered new, but also for the time period between calibrations.

All measurement instruments are prone to drift over time. Unfortunately, from a process view this means there is always some risk that instruments currently in use are actually out of specification. To mitigate this risk, calibration intervals are assigned based on manufacturer specifications, process requirements, equipment history, and acceptable risk requirements.

In most cases the manufacturer's specifications are weighted heaviest when determining suitability of equipment for a process. It can be difficult to make a proper determination of an instrument's expected long-term performance because all of its specifications must be looked at to gain proper insight into its performance. It is not sufficient to look solely at the instrument's accuracy. That usually does not take into account long-term (or one year) drift of the instrument, or may only be applicable to a narrow operating environment. Not properly understanding the performance of an instrument will lead to unpleasant surprises during the next calibration cycle.

Some instrument types and applications tend to be easier to define than others. A high-end digital multimeter used in a laboratory setting can be relatively straightforward. Many manufacturers will state 24-hour, one-week, one-month, and one-year accuracies as well as provide guidance about the affect of temperature on the readings. The behavior of these types of equipment is well known and, in the grand scheme of metrology, relatively predictable. On the other end of the spectrum are instruments such as relative humidity meters. These are used in such a wide variety of operating conditions and environments that it is difficult for manufacturers to specify the right instruments for every application. Every environment is different and could have a different effect on the sensors.

When looking at specifications, always check to see if all the factors are included in the stated accuracy. Many times the stated accuracy is only valid for a specific temperature range, or it may not include the calibration uncertainty, or perhaps the long-term drift is not included. It is important to take a careful look at what is and is not being included in an accuracy specification. If it does not explicitly say that long-term drift is included in the accuracy, it likely is not and must be taken into account separately.

5.4 INSPECTING AGRICULTURAL EQUIPMENT

The purpose of an inspection is to identify whether work equipment can be operated, adjusted and maintained safely – with any deterioration detected and remedied before it results in a health and safety risk. Not all work equipment needs formal inspection to ensure safety and, in many cases, a quick visual check before use will be sufficient. However, inspection is necessary for any equipment where significant risks to health and safety may arise from incorrect installation, reinstallation, deterioration or any other circumstances. The need for inspection and inspection frequencies should be determined through risk assessment.

You should inspect work equipment if your risk assessment identifies any significant risk (for example, of major injury) to operators and others from the equipment's installation or use. The result of the inspection should be recorded and this record should be kept at least until the next inspection of that equipment. Records do not have to be made in writing but, if kept in another form (eg on a computer), these should be held securely and made available upon request by any enforcing authority.

Work equipment that requires inspection should not be used, unless you know the inspection has taken place. Where it leaves your undertaking, or is obtained from another (eg a hire company) it should be accompanied by physical evidence of the last inspection, such as an inspection report or, for smaller items of equipment, some form of tagging, colour coding or labelling system.

5.4.1 Coverage of Inspection

This will depend on type of work equipment, its use and the conditions to which it is exposed. This should be determined through risk assessment and take full account of any manufacturer's recommendations. The advice of others, such as trade associations and consultants, as well as other sources like published advice on health and safety, may also be helpful.

An inspection should concentrate on those safety-related parts which are necessary for the safe operation of work equipment and, in some cases, this may require testing or dismantling. However, not all safety-critical features on a particular item of work equipment may require inspection at the same intervals.

An inspection can vary in its extent, as the following demonstrate:

- quick checks before use (eg electric cable condition on hand-held power tools, functional testing of brakes, lights on mobile machinery)
- weekly checks (eg presence of guarding, function of safety devices, tyre pressures, and the condition of windows, mirrors and CCTV on mobile plant)
- more extensive examinations, undertaken every few months or longer (eg general condition of a ladder, close examination of a safety harness, portable appliance testing)
- Records are not normally required to be made for the simplest pre-use checks.

The use of checklists can assist but these, and the records made, should be tailored to the particular type of work equipment to minimise the burden to what is strictly necessary for safety. Requiring too much detail too often can lead to inspection activity becoming burdensome with the risk of a superficial 'tick box' approach or even, in some cases, the inspection activity ceasing altogether. You only need to inspect what is necessary for safety.

When should work equipment that needs inspection be re-inspected? Work equipment which is exposed to conditions causing deterioration that could result in a dangerous situation should be inspected at suitable intervals, and after every event liable to jeopardise its safety. The frequency of inspection may vary, depending on environmental conditions (eg equipment subject to harsh outdoor conditions is likely to need more frequent inspections than if used in an indoor environment).

The frequency of inspection should be determined through risk assessment, taking account of the manufacturer's recommendations, industry advice and your own experience. It may be appropriate to review the frequency of inspection in the light of your experience. Intervals between inspections can be increased if the inspection history shows negligible deterioration, or shortened where experience shows this is necessary to prevent danger.

5.5 DETECTION OF FAULTS IN AGRICULTURE EQUIPMENT

There are mechanical fault diagnosis many ways, both static-based machinery fault diagnosis method, there are dynamic machinery fault diagnosis method based on which the mechanical vibration failure detection mechanical object of detection, is the mechanical failure detection field a mainstream approach. However, the accuracy of the diagnosis based on the type of fault and the fault identification mechanical vibration fault detection method largely depends on the mechanical vibration signal analysis and processing capabilities. At present, diagnosis of mechanical failure, the vast majority of studies have focused on mechanical vibration signal acquisition and analysis of algorithms, which also includes a large number of the acquired mechanical vibration signals of different transformation or signal conversion other treatment to improve the ability to identify the mechanical vibration signal. However, at present this series of typical mechanical fault diagnosis, fault diagnosis algorithm and the time it was designed, to a large extent depends on statistical or probabilistic analysis, and not very clear fault identification basis. Precisely because of this, mechanical failure detection method currently proposed has strong limitations, that is, for different scenarios, different test objects, and its ability to effect fault detection is often very different.

Agricultural machinery fault recognition system comprising the following modules,

a) monitoring module, the monitoring module operating state of the agricultural machine member into a digital signal and stored;

b) comparing module, comparing module stores a predetermined operating parameter values of the normal operation of agricultural machinery parts, and monitoring module transmitted digital signal with a preset operating parameter values;

c) judging module, judging module determines that the comparison result of the comparison module, whether the category of mechanical failure;

d) Drive module, the determination module determination result of the cloud uploaded to the cloud server module, and find the most preferred solution according to the determination result of the fault;

The present invention can solve the problems of current technology, to improve the fault detection so that the efficiency of agricultural machinery, increased automation.

5.6 MALFUNCTIONS OF AGRICULTURE EQUIPMENTS

5.6.1 Cause of Malfunction

Equipment failure is a common problem for many organizations. Depending on repair costs, total downtime, and subsequent impacts on the production of goods or delivery of services, equipment failure can have devastating and far-reaching effects. Not all equipment failure falls into this category, of course; many malfunctions and broken parts are easily rectified and result in minimal losses. However, without proper planning, maintenance, and monitoring, your next equipment failure could be catastrophic.

In other cases, the equipment may be used with more frequency or for more intensive applications than it was designed to handle. Some examples include:

- ☆ Crane collapses
- ☆ Scaffolding collapses
- ☆ Power tool malfunctions
- ☆ Heavy machinery malfunctions

Equipment malfunction can be caused by a number of factors. It may involve improper handling or use by operator. It could also involve equipment that is poorly maintained or worn out. Sometimes, the equipment is defective as designed.

Improper operation

Equipment operators typically receive in-depth training on appropriate operating procedures, basic troubleshooting, and best practices for safe equipment use relevant to the machines they'll be working with. However, the day inevitably comes when an operator ends up working on a machine that they haven't been adequately trained for. Sometimes this situation arises as a result of short staffing or unexpected absences. Other times emergencies come up that require prompt remediation with available staff–who don't necessarily have the level of expertise that your most experienced operators have.

One solution to these problems is to ensure that you have enough trained operators (ideally, more than you need) to allow for some flexibility and a contingency plan for staff shortage emergencies. If possible, all of your operators should have some training even on every piece of equipment—even assets they don't typically work with.

Most importantly, never allow an operator to use equipment they are not qualified to run. Not only will this help to reduce operational errors, but in some industries, it's imperative for regulatory compliance. The Occupational Safety and Health Administration (OSHA) sets regulations for operator training requirements for certain types of equipment and for general occupational safety. It's up to you, however, to know the regulations applicable to your industry and ensure that you have adequate compliance procedures in place.

Failure to perform preventive maintenance

Most equipment requires regular maintenance for optimal performance, but too often, preventive maintenance is the first task to go when you're short-staffed and overwhelmed. It's easy to brush off regular maintenance when things seem to be running just fine, and many companies work under the assumption that experienced workers will identify impending trouble before equipment fails entirely.

That said, many of the subtle signs of slowing performance or the early stages of failure aren't easily detectable and often go unnoticed, or worse, workers implement temporary fixes (sometimes described as "band-aids") to maintain production in the short-term. Band-aid fixes almost always make the problem worse over time. In other cases, companies simply lack efficient planning methods for ensuring that ongoing maintenance is performed. Tracking equipment and machinery with asset tags can help to keep maintenance schedules on track and equipment operating at maximum operational efficiency.

Preventive maintenance is one ongoing function that should never be allowed to fall by the wayside. Taking care of your equipment with regular tune-ups will extend the usable life of your equipment and machinery, ultimately giving you more use for every dollar. Additionally, preventive maintenance can identify small problems with inexpensive solutions before they become major, costly breakdowns. When you use effective inventory control strategies to ensure that you have the right spare parts in supply for the most common maintenance tasks and malfunctions, downtime for routine maintenance and repairs is minimal. In fact, the U.S. Department of Energy estimates that preventive maintenance results in:

- Up to a 30% reduction in energy and maintenance costs
- 35% to 45% fewer breakdowns

- Reductions in downtime by up to 75%.
- The bottom-line benefits are impossible to deny.
- Preventive maintenance should be a priority for all engaged in equipment operations.

Failure to continuously monitor equipment

Beyond conducting regular preventive maintenance, continuous equipment monitoring is a beneficial process that can result in dramatic reductions in breakdowns and subsequent downtime. Continuous monitoring makes use of sensors to establish baselines and detect subtle changes in factors like vibration, which can be used to predict impending breakdowns and failures, often allowing more time for contingency planning and scheduling downtime for equipment to minimize production interruptions.

Continuous equipment monitoring falls under the umbrella of predictive maintenance, which aims to detect changes in the functional state of equipment. In turn, companies can identify the causes of increased stress on equipment or machinery and adjust workloads and schedules to lessen the load on machinery showing early indicators of impending failure.

Predictive maintenance is different from preventive maintenance in that rather than perform a set of tasks (such as cleaning, oil changes, tune-ups, etc.) on a pre-determined schedule, it evaluates the actual condition and operational status of equipment in real-time, detecting subtle changes that go unnoticed even by the most experienced operators and skilled maintenance professionals.

Equipment can be unpredictable, and eventual failure is inevitable. But by ensuring adequate operator training, planning for appropriate staffing levels based on operator needs, and implementing preventive and predictive maintenance strategies, you'll keep your equipment running at optimal performance and benefit from a longer lifespan from your most valuable assets.

5.6.2 Accidents due to malfunction of machinery

Accidents in agriculture claim many lives each year throughout the country and can be a tremendous challenge to emergency responders for many reasons. The accident may not be discovered and reported immediately, and the farm may be remotely located and have poor access for emergency services. The resulting delayed response times may mean a poor victim outcome. Common machine-related accidents in the workplace are often caused by:

- Defective switch accidents
- Crane accidents
- Bulldozer accidents
- Loader accidents

- ☆ Forklift accidents
- ☆ Defective conveyor belt accidents
- ☆ Safety guard accidents
- ☆ Drill press accidents
- ☆ Factory machine accidents

Agriculture machinery incidents make up a large part of farm-related injuries and may result from the equipment's age, lack of maintenance or missing safety equipment; the operator's inadequate training; failure to read the operator's manual; ignoring safety warnings; or failure to follow safety rules. Any of these factors can result in entanglement or injury.

Agriculture machinery's construction differs from that of highway vehicles from which we are accustomed to extricating victims. Agriculture equipment is heavier and has a different center of gravity; the metal used in its construction may be stronger than that in our rescue tools. Responders must be trained in specific rescue procedures designed for these types of incidences. Agriculture programs are an excellent way to learn or brush up on decision making, stabilization, and machinery extrication skills. Visiting a local agriculture equipment dealer is one good way to learn about the different types of machinery used in farming; the dealer is a valuable resource for your agriculture rescue resource list.

Power take-off (PTO) accidents can be very serious, resulting in life-threatening injuries and even death. A PTO is used to transfer power from the tractor to a variety of implements through a shaft. The stub is the part to which PTO-powered implements are attached on the tractor; typically, the number of splines on the stub will indicate the PTO's speed; usually the more splines, the faster the PTO. The human body is no match for the impressive power of a PTO, which can have speeds of between 540 and 1,000 rpm.

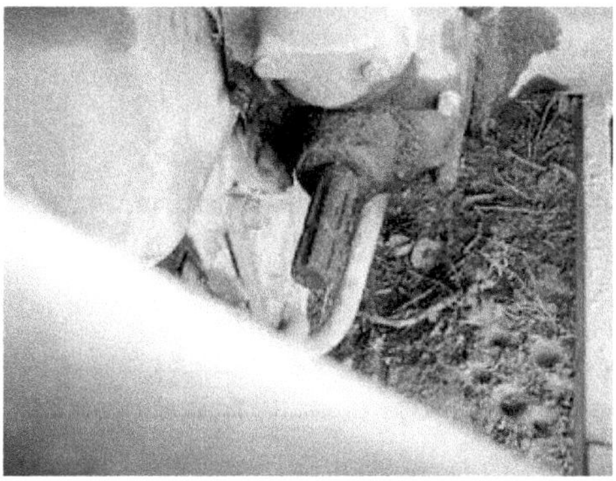

The shaft from the implement is usually connected to the stub by a spring-loaded push pin or spring-loaded collar. On older implements, you may even find a hole designed for a pin to fit into, but normally the pin was lost long ago and something else has been substituted for it.

The PTO shafts will usually be of solid or tubular shaft construction. If cutting the shaft is the most feasible option in an incident, look for the weakest point to make your cut. Exothermic torches are great tools for cutting shafts. The down side is they produce sparks.

Not all tractor or PTO incidents happen on a farm. Look around your response area; farm-type machinery has many applications in urban and industrial settings. Streets and maintenance departments and construction crews may use tractors with front-end loaders, backhoes, scrape blades, and many hydraulic or PTO-driven implements such as spreaders and mowers. Also, golf courses and city parks use agricultural-type machinery in everyday operations. It is essential that we know how to handle these types of emergencies regardless of whether we work in an agriculture area or at a downtown city station.

5.6.3 Injuries and Impact of defective tool

- ☆ The tip of a chisel used as a screwdriver can break and fly off, hitting the user or a nearby worker.
- ☆ The head of a hammer or ax can fly off and strike someone else if the wooden handle is loose, cracked or splintered.
- ☆ Abrasive wheel tools may disintegrate or explode during startup, sending flying fragments at workers.
- ☆ Workers can be hit by pneumatic tool attachments or fasteners being used with the tool.
- ☆ Fuel vapors from liquid fuel tools can burn or explode, injuring workers.
- ☆ Power-actuated tools operate like loaded guns and are so dangerous that they should be used only by specially trained workers.
- ☆ Injuries Caused by Construction Equipment Malfunction

- ☆ A wide range of large equipment and powerful tools are used in the construction industry on a daily basis.
- ☆ When equipment malfunctions, it can cause injury to the worker operating the equipment or to other workers nearby.

Injuries caused by defective or malfunctioning equipment can be serious or catastrophic. They may include:

- ☆ Amputations.
- ☆ Bone fractures.
- ☆ Burns.
- ☆ Traumatic brain injuries (TBI).
- ☆ Severe lacerations.
- ☆ Spinal cord injuries.
- ☆ Crush Injuries.
- ☆ Electrical shock.
- ☆ Eye injuries and loss of vision.

Chapter 6

ADJUSTMENT OF AGRICULTURE EQUIPMENTS

6.1 DEFECTIVE AGRICULTURE EQUIPMENT

Working with heavy machinery and equipment on a farm brings risk. In fact, even the tools require regular maintenance to ensure safety. However, sometimes that is not enough. Faulty parts and defective farm equipment can cause serious and even fatal injuries.

Most farmers understand that agriculture is dangerous. They respect and try to be as careful as possible while working. However, many are injured, and it is beyond their control. These injuries are from defective farm equipment which includes hay balers, plows, tractors, harvesters, shredders and other equipment.

Defective farm equipment seriously injures or kills many people every year. In fact, farm fatalities are on the rise. Defective tractors, mowers, shredders, and other equipment cause debilitating and often lifelong injuries like broken bones, crush injuries, burns, electrocution, amputation and more.

6.1.1 Design Defects

When a company designs equipment without proper safety guards, shutoff switches, and protective enclosures agricultural workers are at risk. This happens all too often like:

- ☆ Pinch points—points at which two objects on the machine make contact, with at least one of them moving in a circle.

- ☆ Wrap points—any exposed, rotating machine component attached at one end to a piece of equipment. These can catch on clothing and limbs, pulling them into the machine.
- ☆ Freewheeling points—after a farmer turns off the equipment, farm equipment parts, such as rotary motor blades and baler flywheels, continue to rotate.
- ☆ Spring points—released springs, which cause injury.
- ☆ Hydraulic lifts and hoses—hydraulic lifts that raise and support heavy objects and assist in steering and braking equipment can fail and crush a worker. This equipment also can eject high-pressure blasts of oil that can burn eyes and skin. Leaky hydraulic hoses can emit toxic fluids that can burn.

6.1.2 Equipment Failures

There are certain unique failure patterns of equipment. Understanding these patterns illustrates why the reduction in maintenance could result in improved performance.

- ☆ Failure Pattern A is known as the bathtub curve and has a high probability of failurewhen the equipment is new, followed by a low level of random failures, and followed by a sharp increase in failures at the end of its life. This pattern accounts for approximately 4% of failures.
- ☆ Failure Pattern B is known as the wear out curve consists of a low level of random
- ☆ failures, followed by a sharp increase in failures at the end of its life. The pattern accounts for approximately 2% of failures.
- ☆ Failure Pattern C is known as the fatigue curve and is characterized by a gradually
- ☆ increasing level of failures over the course of the equipment's life. This pattern accounts for approximately 5% of failures.
- ☆ Failure Pattern D is known as the initial break in curve and starts off with a very low level of failure followed by a sharp rise to a constant level. This pattern accounts for
- ☆ approximately 7% of failures
- ☆ Failure Pattern E is known as the random pattern and is a consistent level of random failures over the life of the equipment with no pronounced increases or decreased related to the life of the equipment. This pattern accounts for approximately 11% of failures.
- ☆ Failure Pattern F is known as the infant mortality curve and shows a high initial failure rate followed by a random level of failures. This pattern accounts for 68% of failures.

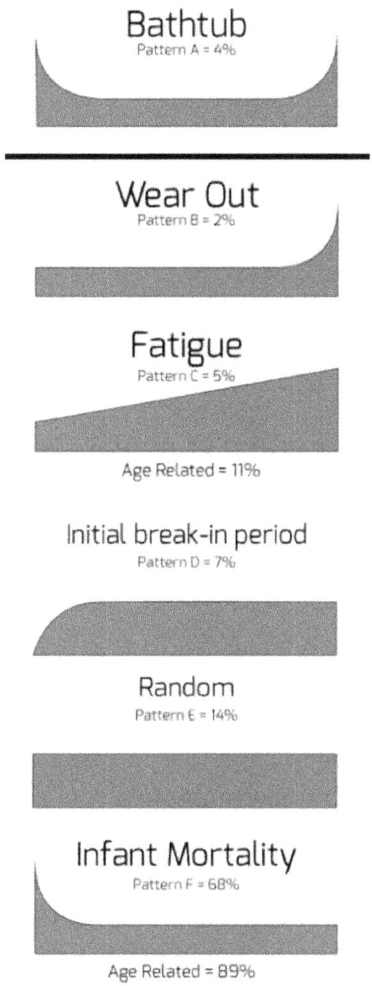

Fig Equipment failure pattern

When looking at the failure patterns, the first three can be group together as the equipment having a defined life, in which the failure rates increase once the equipment has reached a certain age. This age may be time or usage such as hours, widgets produced etc. The failures are usually related to the wear, erosion or corrosion and are often simple components which come into contact with the product. The total of these time based failures only account for 9% of all failures.

The other patterns highlight the fact the during the initial start-up of the equipment is when the majority of failure will occur. This could be due to maintenance induced failures, or manufacturing defects in the components. Once the initial start-up period has passed the failure are random. These patterns account for 86% of failure.

Now, these patterns state that the failures are random in nature, but that does not mean that they failures cannot be predicted or mitigated. It means that overhauls, and replacements conducted at a specific frequency are only effective in 9% of the cases.

In the rest of the failures, the equipment can be monitored and the right time to conduct a replacement, or overhaul is identified based on the condition of the equipment. This is known as Condition Based Maintenance, or Predictive Maintenance or even Don't Fix It Unless It's Broke!

6.2 REPLACEMENT OF DEFECTIVE AGRICULTURE EQUIPMENTS

6.2.1 Calculating Cost

Repair costs occur because of routine maintenance, wear and tear, and accidents. Repair costs for a particular type of machine vary widely from one geographic region to another because of soil type, rocks, terrain, climate, and other conditions. Within a local area, repair costs vary from farm to farm because of different management policies and operator skill.

The best data for estimating repair costs are records of your own past repair expenses. Good records indicate whether a machine has had above or below average repair costs and when major overhauls may be needed. They will also provide information about your maintenance program and your mechanical ability. Without such data, though, repair costs must be estimated from average experience.

The values in Table show the relationship between the sum of all repair costs for a machine and the total hours of use during its lifetime, based on historical repair data. The total accumulated repair costs are calculated as a percent of the current list price of the machine, since repair and maintenance costs usually change at about the same rate as new list prices.

Type of Machinery	Accumulated hours									
	1,000	**2,000**	**3,000**	**4,000**	**5,000**	**6,000**	**7,000**	**8,000**	**9,000**	**10,000**
Two-wheel drive tractor	1%	3%	6%	11%	18%	25%	34%	45%	57%	70%
Four-wheel drive tractor	0%	1%	3%	5%	8%	11%	15%	19%	24%	30%
	200	**400**	**600**	**800**	**1,000**	**1,200**	**1,400**	**1,600**	**1,800**	**2,000**
Moldboard plow	2%	6%	12%	19%	29%	40%	53%	68%	84%	101%
Heavy-duty disk	1%	4%	8%	12%	18%	25%	32%	40%	49%	58%
Tandem disk	1%	4%	8%	12%	18%	25%	32%	40%	49%	58%
Chisel plow	3%	8%	14%	20%	28%	36%	45%	54%	64%	74%
Field cultivator	3%	7%	13%	20%	27%	35%	43%	52%	61%	71%
Harrow	3%	7%	13%	20%	27%	35%	43%	52%	61%	71%
Roller-packer, mulcher	2%	5%	8%	12%	16%	20%	25%	29%	34%	39%
Rotary hoe	2%	6%	11%	17%	23%	30%	37%	44%	52%	61%
Row crop cultivator	0%	2%	6%	10%	17%	25%	36%	48%	62%	78%
	200	**400**	**600**	**800**	**1,000**	**1,200**	**1,400**	**1,600**	**1,800**	**2,000**
Combine heads	0%	2%	4%	8%	14%	21%	30%	41%	54%	69%
Potato harvester	2%	5%	9%	14%	19%	25%	30%	37%	43%	50%
Mower-conditioner	1%	4%	8%	13%	18%	24%	31%	38%	46%	55%
Mower-conditioner (rotary)	1%	3%	6%	10%	16%	23%	31%	41%	52%	64%
Rake	2%	5%	8%	12%	17%	22%	27%	33%	39%	45%
Rectangular baler	1%	4%	9%	15%	23%	32%	42%	54%	66%	80%
Large square baler	1%	2%	4%	7%	10%	14%	18%	23%	29%	35%
Forage harvester (pull)	1%	3%	7%	10%	15%	20%	26%	32%	38%	45%
	300	**600**	**900**	**1,200**	**1,500**	**1,800**	**2,100**	**2,400**	**2,700**	**3,000**
Forage harvester (SP)	0%	1%	2%	4%	7%	10%	13%	17%	22%	27%
Combine (SP)	0%	1%	2%	4%	6%	9%	12%	16%	20%	25%
Windrower (SP)	1%	2%	5%	9%	14%	19%	26%	35%	44%	54%
Cotton picker (SP)	1%	4%	9%	15%	23%	32%	42%	53%	66%	79%
	100	**200**	**300**	**400**	**500**	**600**	**700**	**800**	**900**	**1,000**
Mower (sickle)	1%	3%	6%	10%	14%	19%	25%	31%	38%	46%
Mower (rotary)	0%	2%	4%	7%	11%	16%	22%	28%	36%	44%
Large round baler	1%	2%	5%	8%	12%	17%	23%	29%	36%	43%
Sugar beet harvester	3%	7%	12%	18%	24%	30%	37%	44%	51%	59%
Rotary tiller	0%	1%	3%	6%	9%	13%	18%	23%	29%	36%
Row crop planter	0%	1%	3%	5%	7%	11%	15%	20%	26%	32%
Grain drill	0%	1%	3%	5%	7%	11%	15%	20%	26%	32%
Fertilizer spreader	3%	8%	13%	19%	26%	32%	40%	47%	55%	63%
	200	**400**	**600**	**800**	**1,000**	**1,200**	**1,400**	**1,600**	**1,800**	**2,000**
Boom-type sprayer	5%	12%	21%	31%	41%	52%	63%	76%	88%	101%
Air-carrier sprayer	2%	5%	9%	14%	20%	27%	34%	42%	51%	61%
Bean puller-windrower	2%	5%	9%	14%	20%	27%	34%	42%	51%	61%
Stalk chopper	3%	8%	14%	20%	28%	36%	45%	54%	64%	74%
Forage blower	1%	4%	9%	15%	22%	31%	40%	51%	63%	77%
Wagon	1%	4%	7%	11%	16%	21%	27%	34%	41%	49%
Forage wagon	2%	6%	10%	14%	19%	24%	29%	35%	41%	47%

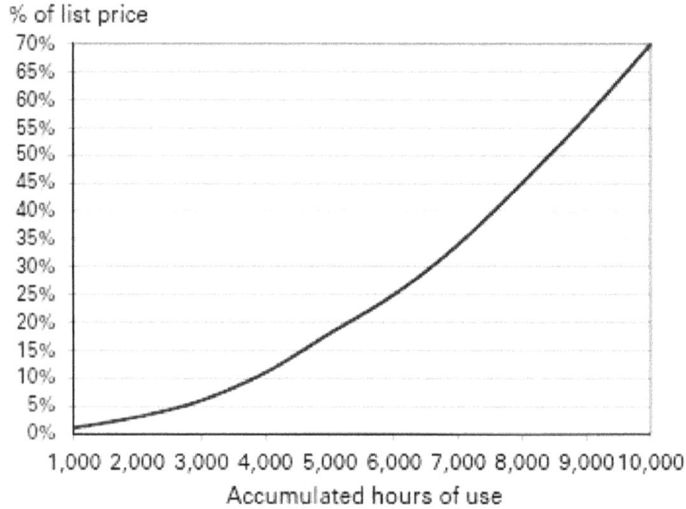

Fig: Shows how repair costs accumulate for two-wheel drive tractors.

Notice the shape of the graph. The slope of the curve increases as the number of hours of use increases. This indicates that repair costs are low early in the life of a machine, but increase rapidly as the machine accumulates more hours of operation.

Because the tractor in the example will be used about 400 hours per year, it will have accumulated about 6,000 hours of operation by the end of its 15-year economic life (400 hours x 15 years = 6,000 hours). According to Table 3, after 6,000 hours of use, total accumulated repair costs will be equal to about 25 percent of its new list price. So, total accumulated repairs can be estimated to be:

Accumulated repairs = 0.25 x $200,000 = $50,000

The average repair cost per hour can be calculated by dividing the total accumulated repair cost by the total accumulated hours:

Repair cost/hour= $50,000 / 6,000 hours = $8.33/hour

Fuel

Fuel costs can be estimated in two ways. AgDM Information File A3-27, Fuel Required for Field Operations (PM 709) lists average fuel use in gallons per acre for many field operations. Those figures can be multiplied by the fuel cost per gallon to calculate the average fuel cost per acre.

For example, if the average amount of diesel fuel required to harvest an acre of corn silage is 3.25 gallons, at a cost of $3.40 per gallon, then the average fuel cost per acre is $11.05.

Average fuel consumption (in gallons per hour) for farm tractors on a year-round basis without reference to any specific implement can also be estimated with these equations:

0.060 x maximum PTO horsepower for gasoline engines

0.044 x maximum PTO horsepower for diesel engines

For our 180-horsepower diesel tractor example:

Average diesel fuel consumption =

0.044 x 180 horsepower = 7.92 gallons/hour

Average fuel cost per hour =

7.92 gallons/hour x $3.40/gallon = $26.93/hour

Lubrication

Surveys indicate that total lubrication costs on most farms average about 15 percent of fuel costs. Therefore, once the fuel cost per hour has been estimated, you can multiply it by 0.15 to estimate total lubrication costs.

For our tractor example, average fuel cost was $26.93 per hour, so average lubrication cost would be:

Lubrication = 0.15 x $26.93 = $4.04/hour

Labor

Because different size machines require different quantities of labor to accomplish such tasks as planting or harvesting, it is important to consider labor costs in machinery analysis. Labor cost is also an important consideration in comparing ownership to custom hiring.

Actual hours of labor usually exceed field machine time by 10 to 20 percent, because of travel and the time required to lubricate and service machines. Consequently, labor costs can be estimated by multiplying the labor wage rate times 1.1 or 1.2. Using a labor value of $15.00 per hour for our tractor example:

Labor cost per hour = $15.00 x 1.1= $16.50

Different wage rates can be used for operations requiring different levels of operator skill.

Total Operating Cost

Repair, fuel, lubrication and labor costs are added to calculate total operating cost. For the tractor example, total operating cost was $55.80 per hour:

Total operating cost = $8.33 + $26.93 + $4.04 + $16.50 =$55.80 per hour

Total Cost

After all costs have been estimated, the total ownership cost per hour can be added to the operating cost per hour to calculate total cost per hour to own and operate the machine. Total cost per hour for our example tractor was:

Total cost = \$40.74 + \$55.80 = \$96.54 per hour

Implement Costs

Costs for implements or attachments that depend on tractor power are estimated in the same way as the example tractor, except that there are no fuel, lubrication, or labor costs involved.

6.2.2 Replacement Strategies for Machinery

A complete line of machinery is one of the largest investments that a farm business can make. Yet, unlike land or buildings, machinery must be constantly monitored, maintained, and eventually replaced. How and when equipment is replaced can mean a difference of thousands of dollars in annual production costs.

Farm record summaries point out some of the differences between the more profitable and the less profitable farms. The table below shows the wide range of machinery cost per acre and machinery investment per acre from a sample of Iowa farms.

Cost Area	High Third	Middle Third	Low Third
Machinery cost/acre	\$116	\$141	\$139
Machinery investment/acre	\$405	\$545	\$559

Costs related to the machinery line clearly had a large effect on whether farms were high profit producers. Many producers give less attention to machinery costs than other cost areas because the cash expenditures are made infrequently. And, once the investment is made, depreciation and interest (except on loans) become noncash costs, and are less visible.

Machinery Costs Change Over Time

One of the difficulties in analyzing machinery costs is that they change over time. Depreciation tends to be great at first, especially for a machine purchased new, but declines over time. Likewise, interest expense is high initially but gradually diminishes. This is true whether the interest cost is cash interest paid on a loan, or an opportunity cost based on revenue foregone by continuing to own a machine year after year.

On the other hand, repair costs may amount to little or nothing when a machine is still under warranty, but eventually increase as parts wear out and maintenance requirements rise. Fuel and lubrication costs usually do not change much over time, although an older engine may eventually lose some degree of fuel efficiency.

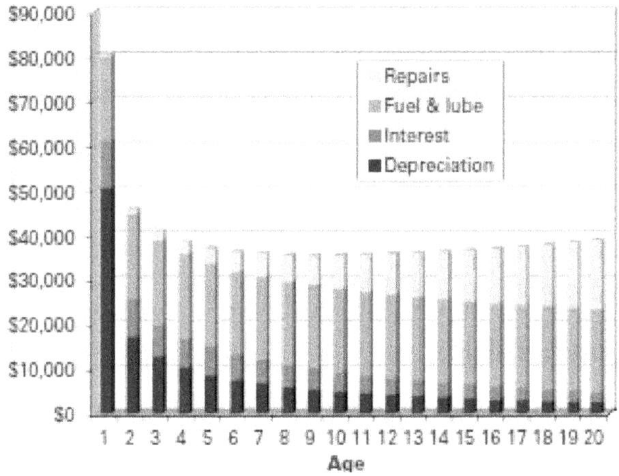

Figure 1 shows how these various costs are projected to change over time, using a 250 horsepower tractor purchased new for $200,000 as an example. Costs are calculated using methods approved by the American Society of Agricultural and Biological Engineers (ASABE). Total costs decline rapidly during the first few years, then rise gradually.

6.3 REPLACING MACHINERY

The decision to replace an item of farm machinery can be made for several reasons.

6.3.1 Cost minimization

The standard rule for minimizing the long-run cost of equipment is to make a change when the annualized total cost of owning and operating the machine begins to increase. In the example shown in Fig, this happens in about the ninth year of ownership. At this point repair costs begin to increase faster than depreciation and interest costs decrease. However, the rate at which total costs rise is often very gradual. In this case, the total cost in year nine is estimated to be $35,706 while the cost by the time the tractor is 15 years old is only $36,896.

Thus, while the rule of increasing total cost can give a general picture of when to replace a particular machine, it cannot give a precise answer. Note that the estimates for repair costs project them to increase gradually over time. In reality, though, repair costs tend to be quite variable from year to year, ranging from only routine maintenance items to a complete overhaul. Being able to anticipate when large repair costs will be needed is a key consideration in deciding when to replace a machine.

6.3.2 Reliability

Besides the standard machinery costs, most operators also consider timeliness costs in their replacement decisions. Timeliness costs occur when crops are not planted or harvested at the optimal time. They can be attributed to losses in yield, such as when corn or soybeans are planted too late to enjoy a full growing season, or a loss of quality, such as when hay or silage is not harvested at its peak nutritional value. If a machine breaks down at a critical time, timeliness costs can be quite high. Timeliness costs are very hard to measure, however, and their importance depends on the weather in any given year. Nevertheless, they should not be ignored, especially in climates where the optimal planting or harvesting period is rather short, and for crops that are particularly sensitive to the effects of weather. Owning machinery that has a high probability of breaking down increases the risk of crop losses.

6.3.3 Pride of ownership

Many farmers take pride in owning and operating new, modern machinery. They may be willing to accept higher long-run costs in return. If the farm business is financially able to bear this cost, there is nothing wrong with "new paint." However, the operators should have a clear idea of how their own machinery costs compare with those of other operations and the scope of their opportunity cost from not having capital invested in other assets.

6.3.4 New technology

In some cases a machine may be in perfectly good working order, but the introduction of new technology has made it obsolete. Newer models may do a better job of harvesting or planting, or operate more efficiently. Care should be taken to distinguish new technology that can increase profits from changes that simply provide more convenience and comfort.

6.3.5 Need for capacity

When the number of acres of crops being produced increases significantly, operators may need to replace machinery with models that have higher capacity to complete planting and harvesting without serious timeliness losses. Likewise, when farm size is reduced, it may be possible to cut costs by downsizing the machinery set.

6.3.6 Farm machinery market

The market for farm machinery is subject to changes in supply and demand, just as for any other product. In particular, the demand for both new and used machinery is strongly affected by ups and downs in the farm economy. The operator who maintains a good capital reserve or borrowing capacity may be able to reduce long-run ownership costs by replacing machinery when dealers have excess inventory and are willing to offer deep discounts to make a sale. When the farm economy is below average, there may be bargains available in used machinery.

6.4 PARTS REPLACEMENT STRATEGIES

There are at least four general strategies that farmers can follow for replacing machinery.

6.4.1 Replace frequently

This approach minimizes the risk of breakdowns and costly repairs by trading key machinery items every few years. Even when repairs occur, they often will be covered by the original warranty. Operators who cover a large number of acres each year and would be severely inconvenienced by extended down time are most likely to follow this strategy. Although this is probably a more expensive approach over the long run, some of the extra costs are offset by fewer timeliness losses, the ability to farm more acres, and less need to invest in repair and maintenance tools and facilities. Operators who trade machinery frequently may find that leasing or rollover ownership plans are more feasible for them than conventional purchase plans. These options are discussed later.

6.4.2 Replace something every year

A second approach is to try to replace one or two pieces of machinery every year. The goal is to spend about the same amount on new equipment each year. This avoids having to make a very large cash outlay in any one year. However, it also could result in replacing machinery before it is really necessary. This strategy often is used by operators who prefer to finance machinery purchases out of their annual cash flow rather than with borrowed money. It works best when the net cash income of the operation is fairly constant from year to year or when significant cash reserves are available.

6.4.3 Replace when cash is available

A third approach is to postpone major machinery purchases until a year when cash income is higher than average. This keeps the machinery purchase from cutting into funds needed for other purposes such as family living and debt servicing. It also helps to level out income for income tax purposes, although rapid depreciation methods and the ability to use income averaging have made this less of a consideration than in previous years. The biggest disadvantage of this strategy is that it is very hard to predict when extra cash will be available. Furthermore, a machine may become seriously unreliable before the business has sufficient funds to replace it.

6.4.4 Keep it forever

Finally, some operators simply hang on to machinery until it reaches the point where it can no longer perform its intended function and is not worth renovating. This may be the least cost approach in the long run, but it runs the risk of a machine failing at a crucial time, or having to arrange financing on short notice. The operator also must be willing to use less than the latest technology. Some older items can be relegated to less critical uses, such as keeping a second planter for a backup unit, or using an older tractor for jobs such as powering an auger or moving wagons. This strategy works best for operators who have considerable flexibility in when they complete key field operations, and who have the skill, patience and facilities to do their own repair and maintenance work.

6.5 CHANGING OF COMPONENTS USING HAND AND POWER TOOLS

6.5.1 Power Tools

Power tools can be extremely dangerous if they are used improperly. Each year, thousands of people are injured or killed by power tool accidents. Common accidents associated with power tools include abrasions, cuts, lacerations, amputations, burns, electrocution, and broken bones. These accidents are often caused by the following:

- Touching the cutting, drilling, or grinding components
- Getting caught in moving parts
- Suffering electrical shock due to improper grounding, equipment defects, or operator misuse
- Being struck by particles that normally eject during operation
- Touching hot tools or work pieces
- Falling in the work area
- Being struck by falling tools

Precautions

When working around power tools, you must wear personal protective equipment and avoid wearing loose clothing or jewelry that could catch in moving machinery. In additional to general shop guidelines, follow these guidelines for working with power tools:

- Use the correct tool for the job. Do not use a tool or attachment for something it was not designed to do.

- ☆ Select the correct bit, blade, cutter, or grinder wheel for the material at hand. This precaution will reduce the chance for an accident and improve the quality of your work.
- ☆ Keep all guards in place. Cover exposed belts, pulleys, gears, and shafts that could cause injury.
- ☆ Always operate tools at the correct speed for the job at hand. Working too slowly can cause an accident just as easily as working too fast.
- ☆ Watch your work when operating power tools. Stop working if something distracts you.
- ☆ Do not rely on strength to perform an operation. The correct tool, blade, and method should not require excessive strength.
- ☆ If undue force is necessary, you may be using the wrong tool or have a dull blade.
- ☆ Before clearing jams or blockages on power tools, disconnect from power source. Do not use your hand to clear jams or blockages, use an appropriate tool.
- ☆ Never reach over equipment while it is running.
- ☆ Never disable or tamper with safety releases or other automatic switches.
- ☆ When the chance for operator injury is great, use a push stick to move material through a machine.
- ☆ Disconnect power tools before performing maintenance or changing components.
- ☆ Keep a firm grip on portable power tools. These tools tend to Keep a firm grip on portable power tools. These tools tend to "get away"
- ☆ Remove chuck keys or adjusting tools prior to operation.
- ☆ Keep bystanders away from moving machinery.
- ☆ Do not operate power tools when you are sick, fatigued, or taking strong medication.
- ☆ When possible, secure work pieces with a clamp or vise to free the hands and minimize the chance of injury. Use a jig for pieces that are unstable or do not lie flat.

Safety

Tools may be powered by electricity, compressed air, hydraulics, belts or chain drives. Power used by tools to do work can posses a tremendous amount of energy that must be controlled by the worker using the tool. Hazards from tool power sources affect not only the person using it, but also to those working close by.

Keep it Sharp

Whether a power tool or a hand tool, if it's designed to cut, keeping the cutting edge sharp is a safety priority. Sharp tools work better and require less force. When sharpening, adjusting or changing a blade, bit or cutting edge, ensure your workers disconnect the tool from the power source by unplugging electrical cords or removing hoses.

Bits, blades and cutting tool edges should be covered while on the shelf or in the toolbox. This not only helps keep the cutting edge from becoming dull, it also prevents hand injuries when a worker reaches for the cutting tool.

Check adjustments

If there are any adjustable parts of a tool, they will most likely be "unadjusted" when a worker needs the tool. Train your people to check all adjustments before using the tool.

Check for damage

Look at handles, tool edges, power cords, hoses & connections, switches, triggers, casings and attachments. Check hand tools for cracks dings and chips. Don't use damaged tools. Generally, hand tools can not be repaired and should be thrown away. Power tools should not be jury rigged to work. Have repairs made only by someone who knows the tool repair craft.

Personal Protective Equipment

Almost all hand or power tool use requires wearing eye protection. If dust, fumes or mists are produced, the proper respirator must also be worn. Using gloves with powered equipment may be a hazard if there is a chance of the glove material getting caught in the equipment or material being worked.

Clothing Hazards

Long draping sleeves, baggy shirts, floppy pant legs are hazardous around powered equipment. Long hair should be pulled back so that it does not hang in front of the ears. All personal jewelry such as rings, necklaces and pendants should be removed before operating powered tools or equipment.

6.5.2 Hand Tool

Hand tools are non-powered. They include anything from axes to wrenches. The greatest hazards posed by hand tools result from misuse and improper maintenance.

Appropriate personal protective equipment such as safety goggles, gloves, etc., should be worn due to hazards that may be encountered while using portable power tools and hand tools.

Around flammable substances, sparks produced by iron and steel hand tools can be a dangerous ignition source. Where this hazard exists, spark-resistant tools made from brass, plastic, aluminum, or wood will provide for safety.

Hazards

Hand tools are tools that are powered manually. Hand tools include anything from axes to wrenches. The greatest hazards posed by hand tools result from misuse and improper maintenance.

- ☆ If a chisel is used as a screwdriver, the tip of the chisel may break and fly off, hitting the user or other employees.
- ☆ If a wooden handle on a tool, such as a hammer or an axe, is loose, splintered, or cracked, the head of the tool may fly off and strike the user or other employees.
- ☆ If the jaws of a wrench are sprung, the wrench might slip.
- ☆ If impact tools such as chisels, wedges, or drift pins have mushroomed heads, the heads might shatter on impact, sending sharp fragments flying toward the user or other employees.

Chapter 7

CLEANING, LUBRICATING AND PERFORMANCE OF AGRICULTURE EQUIPMENTS

7.1 INTRODUCTION

Agriculture is one of the most hazardous sectors. Non-fatal work-related accident rates are 1.7 times higher than the average, and the number of fatal occupational accidents is three times higher than the average. The main causes of accidents include:

- ☆ transport (being run over or vehicle overturns);
- ☆ falling from height (through fragile roofs, from trees etc);
- ☆ struck by moving or falling objects (bales, trees etc);
- ☆ trapped by something collapsing or overturning;
- ☆ livestock related accidents and fatalities;
- ☆ asphyxiation / drowning.

Maintenance influences almost all aspects of farm work, be it in the state of buildings and infrastructure, or the operation of machines and equipment. Farm workers may get harmed during maintenance work. Farm workers may get harmed because of lack of maintenance or as a result of poor maintenance. Maintenance activities in agriculture are various and include the maintenance of both machinery and infrastructure, and the tasks range from oil and filter changes, battery charging and replacement, lubrication, clearing blockages and maintenance of hydraulic systems and tractors to maintenance of roofs and glasshouses, maintenance of silos, slurry tanks, bins and grain tanks or maintenance of electric fences and roads.

Because of the wide variety of tasks, there are many different hazards involved, including:

- ☆ Mechanical hazards related to the maintenance of machinery, such as crushing, entanglement and high-pressure fluid injection
- ☆ Electrical hazards when working with defective equipment or during maintenance of electrical installations and equipment, or repair of broken electric fences
- ☆ Thermal hazards related to the use of welding or heating equipment during maintenance, or maintenance of equipment with hot surfaces or operating fluids
- ☆ Chemical hazards related to the use of dangerous substances during maintenance, or maintenance of equipment containing dangerous substances
- ☆ Fire or explosion hazard during maintenance of facilities or equipment containing dangerous and explosive substances such as tanks, bins and silos, or fuel tanks
- ☆ Biological hazards during maintenance of installations contaminated by biological agents, slurry tanks, ditches and sewage infrastructure
- ☆ Ergonomic hazards, such as awkward postures, poorly designed tools
- ☆ Hazards related to working in confined spaces
- ☆ Falls from height, slips, trips

7.2 MAINTENANCE

The key to keep equipment running and staying at peak performance is no secret. Every successful technician knows it's all about maintenance of agricultural tools and equipment. They know the reasons why agriculture machines should be maintained and the value in service of farm equipment.

7.2.1 Types of maintenance

Normally there are 5 types of maintenance which are differentiated by the nature of the tasks that they include:

- ☆ Corrective maintenance: The set of tasks is destined to correct the defects to be found in the different equipment and that are communicated to the maintenance department by users of the same equipment.
- ☆ Preventive Maintenance: Its mission is to maintain a level of certain service on equipment, programming the interventions of their vulnerabilities in the most opportune time. It is used to be a systematic character, that is, the equipment is inspected even if it has not given any symptoms of having a problem.
- ☆ Predictive Maintenance: It pursues constantly know and report the status and operational capacity of the installations by knowing the values of certain variables, which represent such state and operational ability. To apply this maintenance, it is necessary to identify physical variables (temperature, vibration, power consumption, etc.). Which variation is indicative of problems that may be appearing on the equipment. This maintenance it is the most technical, since it requires advanced technical resources, and at times of strong mathematical, physical and / or technical knowledge.
- ☆ Zero Hours Maintenance (Overhaul): The set of tasks whose goal is to review the equipment at scheduled intervals before appearing any failure, either when the reliability of the equipment has decreased considerably so it is risky to make forecasts of production capacity . This review is based on leaving the equipment to zero hours of operation, that is, as if the equipment were new. These reviews will replace or repair all items subject to wear. The aim is to ensure, with high probability, a good working time fixed in advance.
- ☆ Periodic maintenance (Time Based Maintenance TBM): the basic maintenance of equipment made by the users of it. It consists of a series of elementary tasks (data collections, visual inspections, cleaning, lubrication, retightening screws,...) for which no extensive training is necessary, but perhaps only a brief training. This type of maintenance is the based on TPM (Total Productive Maintenance).

Proper maintenance of agriculture machineries reduces the expenses needed to keep a farm running at peak efficiency, extends the life of tools and equipment and prevents sudden breakdowns. Spending a lot of time and money maintaining farm equipment might seem unnecessary to an outsider, however, every farmer who's been in the agriculture business and intends on remaining on a profitable farm knows the cost of maintenance is far less than paying for even more expensive repairs.

7.2.2 Routine maintenance of Agriculture equipment

Heavy vehicles and mobile equipment are indispensable to many industrial activities, from construction to railroad transportation. Various types of equipment move materials, till land, lift beams, and dig earth to pave the way for development and production. Heavy vehicle and mobile equipment service technicians and mechanics repair and maintain engines and hydraulic, transmission, and electrical systems for this equipment. Farm machinery, cranes, bulldozers, and railcars are all examples of heavy vehicles that require such service.

Service technicians perform routine maintenance checks on agricultural, industrial, construction, and rail equipment. They service fuel, brake, and transmission systems to ensure peak performance, safety, and longevity of the equipment. Maintenance checks and comments from equipment operators usually alert technicians to problems. After locating the problem, these technicians rely on their training and experience to use the best possible technique to solve it.

With many types of modern equipment, technicians can use diagnostic computers to diagnose components needing adjustment or repair. If necessary, they may partially dismantle affected components to examine parts for damage or excessive wear. Then, using hand-held tools, they repair, replace, clean, and lubricate parts as necessary. In some cases, technicians re-calibrate systems by typing codes into the onboard computer. After reassembling the component and testing it for safety, they put it back into the equipment and return the equipment to the field.

Many types of heavy and mobile equipment use hydraulic systems to raise and lower movable parts. When hydraulic components malfunction, technicians examine them for fluid leaks, ruptured hoses, or worn gaskets on fluid reservoirs. Occasionally, the equipment requires extensive repairs, as when a defective hydraulic pump needs replacing.

Service technicians diagnose electrical problems and adjust or replace defective components. They also disassemble and repair undercarriages and track assemblies. Occasionally, technicians weld broken equipment frames and structural parts, using electric or gas welders.

Technicians use a variety of tools in their work: power tools, such as pneumatic wrenches to remove bolts quickly, machine tools, like lathes and grinding machines, to rebuild brakes, welding and flame-cutting equipment to remove and repair exhaust systems, and jacks and hoists to lift and move large parts. Service technicians also use common hand tools—screwdrivers, pliers, and wrenches—to work on small parts and to get at hard-to-reach places. They may use a variety of computerized testing equipment to pinpoint and analyze malfunctions in electrical systems and other essential systems. Tachometers and dynamometers, for example, can be used to locate engine malfunctions. Service technicians also use ohmmeters, ammeters, and voltmeters when working on electrical systems. Employers typically furnish expensive power tools, computerized engine analyzers, and other diagnostic equipment, but hand tools are normally accumulated with experience, and many experienced technicians have thousands of dollars invested in them.

It is common for technicians in large shops to specialize in one or two types of repair. For example, a shop may have individual specialists in major engine repair, transmission work, electrical systems, and suspension or brake systems. Technicians in smaller shops, on the other hand, generally perform multiple functions.

Technicians also specialize in types of equipment. Mobile heavy equipment mechanics and service technicians, for example, keep construction and surface mining equipment, such as bulldozers, cranes, graders, and excavators in working order. Typically, these workers are employed by equipment wholesale distribution and leasing firms, large construction and mining companies, local and Federal governments, and other organizations operating and maintaining heavy machinery and equipment fleets. Service technicians employed by the Federal Government may work on tanks and other armored military equipment.

Farm equipment mechanics service, maintain, and repair farm equipment, as well as smaller lawn and garden tractors sold to homeowners. What once was a general repairer's job around the farm has evolved into a specialized technical career. Farmers have increasingly turned to farm equipment dealers to service and repair their equipment because the machinery has grown in complexity. Modern equipment uses more computers, electronics, and hydraulics, making it difficult to perform repairs without specialized training and tools.

7.2.3 Tips for large machinery maintenance

Heavy machinery, especially Mining, Industrial or Farming Equipment, requires constant maintenance to keep it in good working order. Conversely, poorly maintained large machinery equipment runs inefficiently. Breakdowns are costly and safety is also an important consideration.

1. Stay on top of large machinery

- ☆ Many types of large machinery have multiple operators. One of the ongoing inspections on any checklist should be overseeing the correct operation of the equipment.
- ☆ Large machinery should be inspected as soon as it is purchased. Operator training is usually done at that point, but training needs to be kept up. Employees come and go, skills become rusty and poor operation leads to breakdowns.
- ☆ Operator manuals can be revised for the specific work situation. They can be rewritten in simpler language. A short manual can be provided to each operator for easy reference. And, if you operate in a paperless environment, you can rest assured operators use the most current version of each manual.
- ☆ One other note is to identify best practices, which can then be applied to other facilities or geographic locations. The knowledge you learn about how to maintain your equipment can become quite valuable – be sure to best leverage this important knowledge and use it at every applicable location.

2. Add and test lubricants frequently

- ☆ Lubricants reduce friction around any moving part. A schedule of good lubrication maintenance extends the life of large machinery equipment and parts.
- ☆ Lubrication is one of the first and most important of maintenance checks. Look for signs of excess oil or grease build-up on pistons. Check for leaks around oil seals.
- ☆ Be sure to use the right lubricant. There are specific kinds of oil and grease for every component. Check the manufacturer's recommendations.
- ☆ Getting the lubricants checked is a good way to diagnose problems with large machinery. Experts analyze particles in the used oil. The makeup of any contaminants will indicate which part may be suffering from wear or breakdown.

3. Check for signs of wear

Vibration, shock, high temperatures, friction and age all contribute to the breakdown of parts in heavy machinery.

- ☆ Vibration can come from gears and belts that are out of alignment
- ☆ Shock can come from accidents and from poor operator technique
- ☆ High temperatures can come from extended use, friction, poor lubrication and worn parts, among other reasons
- ☆ Age affects many key components. Over time, belts will warp. Seals will dry and crack. Bolts will loosen and stretch out of shape. Age is a factor to monitor in equipment.

Should you discover wear and tear on any moving parts within your heavy equipment, be sure to quickly perform the necessary replacement of any worn parts.

4. Keep large machinery clean, and maintain a clean environment

There are many seals and filters in place on heavy machinery to keep working parts clean and free of contamination. Seals should be inspected regularly to make sure they're in good condition. Filters should be inspected and changed regularly. Breathers should be kept clean to avoid creating a vacuum in the cab which will suck contaminants into the cab. The electronics in the cab are susceptible to breakdown if contaminated. This impacts the clutch, for example.

Large machinery should be stored in a shed or other building if at all possible. Exposure to wind and weather can lead to rust and rot. The machinery should be run periodically if it is not in use.

5. Have a maintenance and repair schedule, and keep good records

Fluids, tires, tracks and electrical systems are among the components that have to be checked regularly for preventive maintenance. Know what needs to be inspected and when. Here are some examples.

- ☆ Power transmissions have many moving parts that need to be maintained in top condition. Gearboxes need to be checked for lubrication, vibration and damage to parts.
- ☆ Friction materials, seals, gaskets and bearings all need to be inspected for wear and replaced. Gears and shafts usually last a long time and don't need to be replaced often, if at all.
- ☆ Drive train components need constant monitoring. Check pulleys and v-belts on CVT transmissions for alignment and wear. Check sprockets for correct meshing with chains and for breaks.
- ☆ Test the oil to diagnose problems. Change filters frequently.
- ☆ Bearings keep great amounts of force running smoothly and are vital to large machinery performance. Check bearing lubrication often. Maintaining bearings well extends their life.
- ☆ Lubricate gears frequently.
- ☆ Do a seal check to prevent bearing raceway contamination.
- ☆ Run torque checks on the bolts. Bolts can elongate and creep over time.

7.3 LUBRICATING AGRICULTURE EQUIPMENTS

7.3.1 Introduction

Lubrication is simply the use of a material to improve the smoothness if movement of one surface over another; the material which is used in this way is called a lubricant. Lubricants are usually liquids or semi-liquids, but may be solids or gases or any combination of solids, liquids, and gases.

The smoothness of movement is improved by reducing friction. This is not, however, always the case, and there may be situations in which it is more important to maintain steady friction than to obtain the lowest possible friction. In addition to reducing or controlling friction, lubricants are usually expected to reduce wear and often to prevent overheating and corrosion. Lubricants are usually divided into four basic classes.

(a) Oils: A general term used to cover all liquid lubricants, whether they are mineral oils, natural oils, synthetics, emulsions, or even process fluids.

(b) Greases: Technically these are oils, which contain a thickening agent to make them semi-solid. It is convenient, however, to include the anti-seize pastes and the semi-fluid greases under the same heading.

(c) Dry lubricants: These include any lubricants, which are used in solid form, and may be bulky solids, paint-like coatings, or loose powders.

(d) Gases: The gas usually used in gas bearings is air, but any gas can be used which will not attack the bearings, or itself decompose.

The advantages and disadvantages of oils stem from their ability to flow easily. Thus, on the credit side, it is very easy to pour them from a container, to feed them into a bearing by dripping, splashing or pumping, and to drain them out of a machine when no longer fir for use. Other advantages are the cooling of a bearing by carrying away heat, and cleaning it by removing debris.

The behavior of greases is very similar to that of oils, but the former are used where the advantages of easy flow are outweighed by the disadvantages. Thus grease do not easily leak out of a machine, or container, do not migrate away, and will form an effective seal against contaminants.

The advantages and disadvantages of solid lubricants are rather like the extremes for greases, where the lubricant will not flow at all. Similarly, the advantages and disadvantages of gas lubricants are like the extremes of oils, where the flow properties are almost too good.

(a) OILS

Mineral Oil

- ☆ Mineral oils are easily the most widely used lubricating oils. Consequently, they are often the standard with which other oils are compared. It may be helpful to consider first what mineral oils are.
- ☆ Mineral oils generally mean oils obtained from petroleum, although they can also be obtained from similar sources, such as oil shales and tar-stands. The mineral oils used for lubrication were originally just the fractions, obtained by distilling petroleum, which had a suitable viscosity for lubrication.

SYNTHETIC AND NATURAL OILS AND EMULSIONS

- ☆ Many of the alternative types of oil are synthetic. They are manufactured from various feedstocks by chemical processes. There are different types of synthetic oil and are very different from each other in their performance and properties.

- ☆ The widely-used synthetic oils include hydrocarbons, di-esters, polyol esters, phosphate ester, silicones, polyglycols, and- to a much smaller extent- polyphenyl ethers and perfluoroalkyl polyethers.

(b) GREASES

- ☆ Greases are not simply very viscous lubricating oils. They are in fact mixtures of lubricating oils and thickeners. The thickeners are dispersed in lubricating oils in order to produce a stable and colloidal structure or gel. Thus, a grease consists of oil constrained by minute thickener fibres.
- ☆ The most widespread application of greases is as low-maintenance, semi-permanent lubricants in rolling contact bearings and some gears. The grease may be packed into a bearing or gear set and left for a period of several months or longer before being replaced. Inaccessible wearing contacts, such as are found on caterpillar track assemblies or in agricultural machinery, are conveniently lubricated by this means,
- ☆ Grease is unable to remove hear by convection as oil does, so unlike oil, it is not effective as a cooling agent. It also cannot be used at speeds as high as oil because frictional drag would cause overheating. The lifetime of a grease in service is often determined by the eventual loss of the semi-solid consistency to become either a liquid or a hard deposit.

(c) SOLID LUBRICANTS

- ☆ Solid lubrication is simply the lubrication of two surfaces in moving contact by means of solid materials interposed between them.
- ☆ Depending upon the nature of the two surfaces, a wide variety of solid materials can reduce the friction and prevent seizure. For example, dust, sand, or gravel on the surface of a road can cause vehicles to skid because they decrease friction between tyres and the road surface.

7.3.2 Lubrication of Agriculture Machinery Parts

If you work in agriculture, you know how important your equipment is to productivity. Whether you are using tractors, bailers, trucks, air compressors, vacuum pumps or other agricultural machinery, keeping equipment running during critical periods is imperative. That can be difficult because of the harsh environmental and operating conditions – such as cold, heat, dirt and water that can be detrimental to a lubricant's performance in your equipment.

- ☆ A good lubrication reliability program starts with having a thorough knowledge of the application and equipment. In agriculture, early starts can be in very cold temperatures, while daytime weather can become extremely hot. Irrigation exposes bearings and gearboxes to water. Deep plowing places heavy loads on tractors.
- ☆ In addition to maintaining the proper lubricant film to reduce wear, frequent lubrication acts as a means to purge bushing and bearings of contaminants. If rock dust, dirt, sand and water are allowed to work their way into these components, they will form a "grinding compound" that reduces bearing life substantially.
- ☆ To properly lubricate manually, it normally takes about 30 minutes per machine and maintenance cycle. Failure to properly lubricate each lubrication point on every machine can have a negative impact on schedules, maintenance costs and running time performance.

7.4 PERFORMANCE OF AGRICULTURE EQUIPMENTS

7.4.1 Factors determines the performance of agriculture equipments

The performance of a machine system is profitable only when it can add value to products and processes beyond the systems cost of operation. A minimum cost would appear to be an optimum economic goal, but overall profit maximization is the true goal of business and on the farm this does not necessarily occur with a minimum cost system of operations. Good machinery management requires that the individual operations in a machine system must be adjusted and combined in a manner such that their overall performance returns the greatest profit to the farm business.

Measures of agricultural machine performance are the rate and quality at which the operations are accomplished. Rate is an important measure because few industries require such timely operations as agriculture, with its sensitivity to season and bad weather. As most of agricultural materials are fragile and many are perishable, the amount of product damage or reduction in product quality due to a machine operation is another important measure of machine performance. Quality as well as quantity must be considered when evaluating machine performance.

There are certain basic principles for predicting performance or getting the best out of existing machinery which includes:

- ☆ Field performance
- ☆ Pattern performance
- ☆ Power performance
- ☆ Quality performance
- ☆ Operator performance

Field Performance

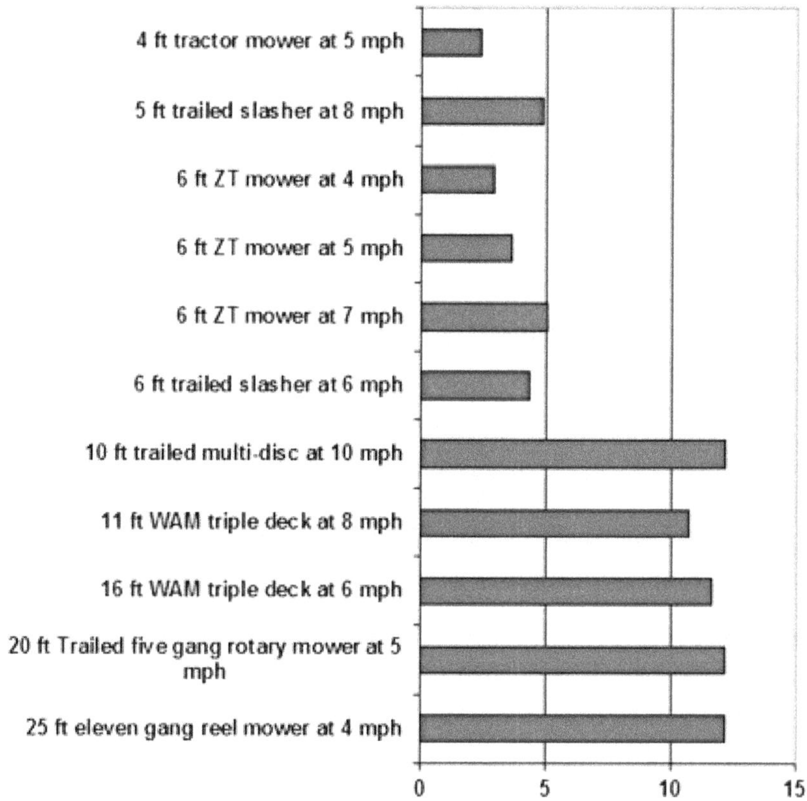

Fig: graph compares the work rate of mowing operations where bottom axis is field capacity in acres per hour.

You can rapidly calculate how big an area you can cover with a simple rule known as:

- ☆ Theoretical Field Capacity (FC)
- ☆ Acres per hour = implement width (in feet) X speed (in mph) ÷ 8.25. Or, in metric units,
- ☆ Theoretical FC; hectares an hour = implement width (m) X speed (km/h) ÷ 10).

- ☆ This is the field capacity of any piece of equipment of given width with the tractor operated at the selected speed.
- ☆ In imperial units, the FC divisor is 8.25, but making it 10 allows for time lost in turns, operator breaks etc; in other words a time loss of about 17.5 per cent short of the theoretical capacity.
- ☆ According to this rule then a wider and faster machine will have the higher FC — unless there are conditions that would interfere with straightforward work.
- ☆ Actual field performance falls short if the Theoretical by the Field Efficiency Factor: Actual Field Capacity (AFC) = Theoretical FC X Field Efficiency.
- ☆ Field efficiency (FE) = Actual Field Capacity X 100 ÷ Theoretical Field Capacity

Some notional FE figures:

- ☆ Mowing FE = 80 - 90%
- ☆ Cultivatin FE = 68-80%
- ☆ Sprayin FE = 55-80%
- ☆ Earthmoving FE = 80%

Material Capacity is relevant to such activities as moving dirt, baling hay or harvesting crops:

- ☆ Tonnes of product per hour = Yield (tonnes/acre) X AFC (acres/hour)
- ☆ Also, Productivity = Load per cycle X cycles per hour
- ☆ Finally, Unit Cost, in dollars per tonne = dollars per hour ÷ tonnes per hour

Pattern Performance

- ☆ Closely connected with Field Efficiency is Pattern Performance or Pattern Efficiency.
- ☆ There are many ways that you can work over an area.
- ☆ What about turning at the corners? At least eight patterns are used in broadacre farming and most of them are pertinent to compact equipment operations as well.

Field operating patterns

- ☆ Certain field operating patterns are more time-efficient than others, particularly those that execute the fastest turns.
- ☆ There are two main pattern types. Pattern type A or' circuitous patterns', sometimes called 'racecourse' or 'round and round' and pattern type B or 'working lands' and 'headland' patterns.

Pattern A: 'Round and Round'

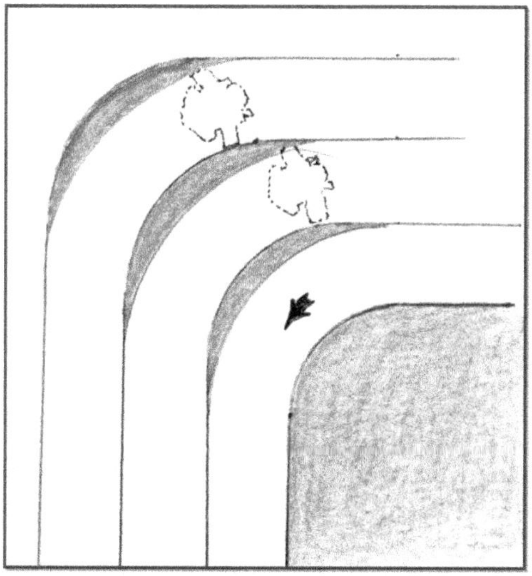

Driving round and round is simple enough, but how about the corners? The two simplest types of turns on a rectangular course are:

- ☆ Ninety-degree continuous turns: These are plain no-nonsense turns but they leave crescent-shaped pieces unworked; the size of crescents depends on the turning radius and width of the implement or attachment.
- ☆ Loop turns: Doing a loop at each corner on the other hand makes a neat cusp and 'cuts out' the corners at each turn so that no further work is needed there. The penalty is turn time that can eat up useful work time, especially in a short or squat field.

There are several other types of fancy turns that are possible but those are more complex and time consuming.

They depend on factors such as whether the attachment is towed, rear- or front-mounted; and whether there is a predetermined operating pattern needed — as there is with a side-mounted sickle bar mower or a plough that throws to one side only.

Turning consumes a lot of otherwise productive time, and the more so the smaller the paddock.

Pattern B: Working Lands and Headland Patterns

- ☆ Working 'Lands' refers to the practice of dividing an area or a field that has straight sides into sub-areas and individually working those sub-areas.
- ☆ That will inevitably call for space to execute the continuous 180 degree turns at the end of each trip.

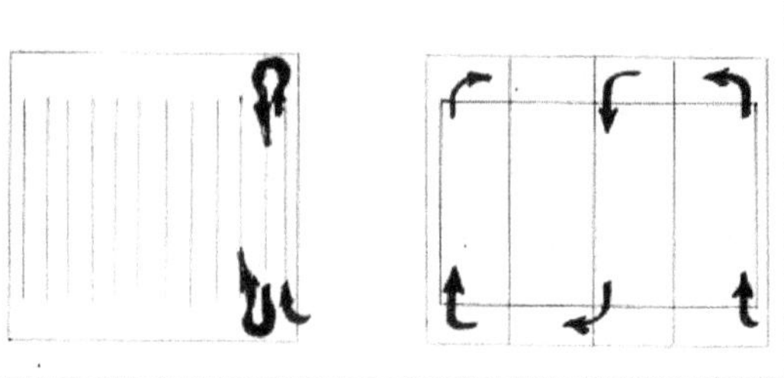

- ☆ Those turn-strips are 'headlands' which will need subsequent processing to tidy up.
- ☆ Headland pattern work is more interesting/challenging for the operator than round and round.
- ☆ For turf striping a sports field, premeditated headland working does leave a nice pattern; the wider the lands, the greater the pattern efficiency.
- ☆ There are several variants on working in lands and turn patterns that have their own advantages and drawbacks.

Power Performance

- ☆ This is measured by the mechanical efficiency of powered mobile equipment to deliver power from the engine flywheel down to where it's needed – to the ground, or through PTOs or hydraulic outlets.
- ☆ In reality most tractor work is done on the go, so that brings the focus on traction: how much power can be delivered to the ground? A brief overview of a big topic follows.
- ☆ Wheeled tractors: If you're working on rough, undulating ground, like towing a trailed brush hog, slashing weeds, needs a tractor with ample power to be able to push on over the hills as well as drive the implement.
- ☆ Where traction is concerned, tyre selection and tyre management are important.

Tyres

- ☆ Tractor tyres are either lugged or non-lugged. The lugs are for traction and should always be mounted with the v-pattern pointing downwards at the front of a traction tyre.

- ☆ The only exception would be for very long raiding, then reversing the tyres will extend lug life.
- ☆ Tractor drive tyres are designated with an R prefix. Non-driving front tyres are often ribbed to aid steering and are designated by an F prefix, and off-roads by an E prefix.
- ☆ Tyres have either bias ply or radial ply inside the rubber carcase. The number of ply inside the tyre dictate the load rating and service work the tyre can withstand.
- ☆ Skid-steer loaders and stake-proof forestry tractors are equipped with tyres with as much as 16-ply.
- ☆ Radials can be operated at lower tyre pressures, which give better traction and a softer ride, but they cost more.
- ☆ Tyre traction depends on the weight on the individual tyre and soil conditions. The load the tyre can carry in turn is governed by the tyre pressure and speed; those have certain limits which are usually inscribed on the tyre.
- ☆ A certain amount of slippage is necessary for any traction and pull; for wheeled tractors on drawbar work that could be around 12 to 15 per cent wheel-slip, but only 3 per cent slip for rubber tracks.
- ☆ Ballasting the tyres or tractor for optimal drawbar pull and balance is complex; for example, there is a different ballast for road work than for field working.
- ☆ Frequently, surface compaction and marking of greens are of concern. For those reasons special flotation tyres and tread patterns are available for greens-keeping tractors.

Quality Performance

There are two facets to the tractor systems' quality performance:

- ☆ The job that the attachments do or what sort of finish they leave
- ☆ The reliability of the tractor and attachments to complete the job

Mowing finish for example can be measured by the appearance of the turf after passage of the mower is there any scalping or unevenness of cut? How about striping? Tractor reliability depends on whether the machine is new or used. For a new machine, initial tractor quality out of the factory.

Reliability factors also include how many hours are on the meter, how the tractor and attachments are used or abused in service and how faithfully service and maintenance are managed in order to keep the machine system operable with minimum downtime.

Breakdowns are inevitable sooner or later. The probability and extent of a breakdown is a combination of many factors — including the operator.

Operator Performance

- ☆ A professional farmer might manage an operation with considerable skill and knowledge of machine performance.
- ☆ But with a careless or inexperienced driver on the machine, overall system performance will be low and profitability will suffer.
- ☆ A manager has certain obligations by law to provide the safest reasonable environment and provide employee training in effective use and safety.
- ☆ Tractors and machines are dangerous. A long day's work behind the wheel, the noise and the need to be constantly alert is fatiguing, apart from any physical effort involved.
- ☆ Modern tractors have sophisticated controls; the more complex, the greater the need for sharpness, knowledge and skill in operation.
- ☆ There's a difference between a driver and an operator. Just how skilful an operator will become depends so much on ability, experience, training, motivation, aptitude, health and alertness.

Ergonomics

A comfortable tractor will enable the operator to stay at it for more hours or get more out of the set-up in the day. Ergonomics has to do broadly with:

- ☆ Comfort – and the operator's platform and controls layout
- ☆ Lowest reasonable noise and vibration levels
- ☆ Facility with the controls
- ☆ Operator perceptions (and even peer pressure)
- ☆ Work organisation
- ☆ Implementing safe work practices.

Look for ergonomically-designed tractor features and use them to operate effectively and safely.

7.4.2 Agricultural Machinery Management

Tractor performance

Power losses occur when transmitting engine power through the drive wheels, the PTO shaft, and the hydraulic system. If Flywheel power is known, multiply it by 0.9 to estimate PTO power.

Drawbar power is the power transferred through the drive wheels or tracks to move the tractor and implement. You can estimate Drawbar power by multiplying the PTO power by the appropriate value from the table below:

Tractor Type	Concrete	Firm Soil	Tilled Soil	Soft Soil
2 wheel drive	0.87	0.72	0.67	0.55
Front wheel assist	0.87	0.77	0.73	0.65
4 wheel drive	0.88	0.78	0.75	0.70
Tracks	0.88	0.79	0.80	0.78

Slippage of drive wheels on soil surfaces is a power loss. Travel reduction, or slip (s), is calculated from:

S = ((An - A1)/An) * 100

where:

- ☆ S is slip, percent;
- ☆ An is the advance under no load conditions per wheel or track revolution, m(ft);
- ☆ A1 is the advance under actual load conditions per wheel or track revolution, m(ft).

Tractive efficiency (TE) is the ratio of drawbar power to axle power and can be estimated when slip is known. TE (and hence tire efficiency) of a wheel can be predicted using a series of equations that take into account tire dimensions, soil conditions, slip, etc. Tractor performance is calculated by summing the individual wheel performances. Maximum TE is obtained with slip ranges of:

- ☆ 4-8% for concrete;
- ☆ 8-10% for firm soil;
- ☆ 11-13% for tilled soil;
- ☆ 14-16% for soft soils and sands

Implement (machine) power requirements

Draft (D) is the total force parallel to the direction of travel that is required to pull the implement. Both functional draft (soil and crop resistance) and draft required to overcome rolling resistance of the implement are included.

Table 1- Craft parameters and an expected range in drafts estimated by the model parameters for tillage and seeding implements.

Implement	SI Units: Width units	SI Units: Machine Parameters A	B	C	English Units: Width units	English Units: Machine Parameters A	B	C	Soil Parameters F_1	F_2	F_3	Range ±%
MAJOR TILLAGE TOOLS												
Subsoiler/Manure injector												
narrow point	tools	226	0.0	1.8	tools	120	0.0	2.7	1.0	0.70	0.45	50
30 cm winged point	tools	294	0.0	2.4	tools	167	0.0	3.5	1.0	0.70	0.45	50
Moldboard Plow	m	652	0.0	5.1	ft	113	0.0	2.3	1.0	0.70	0.45	40
Chisel Plow												
5 cm straight point	tools	91	5.4	0.0	tools	52	4.9	0.0	1.0	0.85	0.65	50
7.5 cm shovel/35 cm sweep	tools	107	6.3	0.0	tools	61	5.8	0.0	1.0	0.85	0.65	50
10 cm twisted shovel	tools	123	7.3	0.0	tools	70	6.7	0.0	1.0	0.85	0.65	50
Sweep Plow												
primary tillage	m	390	19.0	0.0	ft	60	5.2	0.0	1.0	0.85	0.65	45
secondary tillage	m	273	13.3	0.0	ft	42	3.7	0.0	1.0	0.85	0.65	35
Disk Harrow, Tandem												
primary tillage	m	309	16.0	0.0	ft	53	4.6	0.0	1.0	0.88	0.78	50
secondary tillage	m	216	11.2	0.0	ft	37	3.2	0.0	1.0	0.88	0.78	30
Disk Harrow, Offset												
primary tillage	m	364	18.8	0.0	ft	62	5.4	0.0	1.0	0.88	0.78	50
secondary tillage	m	254	13.2	0.0	ft	44	3.8	0.0	1.0	0.88	0.78	30
Disk Gang, Single												
primary tillage	m	124	6.4	0.0	ft	21	1.9	0.0	1.0	0.88	0.78	25
secondary tillage	m	86	4.5	0.0	ft	15	1.3	0.0	1.0	0.88	0.78	20
Coulters												
smooth or ripple	tools	55	2.7	0.0	tools	31	2.5	0.0	1.0	0.88	0.78	25
bubble or flute	tools	66	3.3	0.0	tools	37	3.0	0.0	1.0	0.88	0.78	25
Field Cultivator												
primary tillage	tools	46	2.8	0.0	tools	26	2.5	0.0	1.0	0.85	0.65	30
secondary tillage	tools	32	1.9	0.0	tools	18	1.8	0.0	1.0	0.85	0.65	25
Row Crop Cultivator												
S-tine	rows	140	7.0	0.0	rows	80	6.4	0.0	1.0	0.85	0.65	15
C-shank	rows	260	13.0	0.0	rows	148	11.9	0.0	1.0	0.85	0.65	15
No-till	rows	435	21.8	0.0	rows	248	19.9	0.0	1.0	0.85	0.65	20
Rod Weeder	m	210	10.7	0.0	ft	37	3.0	0.0	1.0	0.85	0.65	25
Disk-Bedder	rows	185	9.5	0.0	rows	106	8.7	0.0	1.0	0.88	0.78	40
MINOR TILLAGE TOOLS												
Rotary Hoe	m	600	0.0	0.0	ft	41	0.0	0.0	1.0	1.0	1.0	30
Coil Tine Harrow	m	250	0.0	0.0	ft	17	0.0	0.0	1.0	1.0	1.0	20
Spike Tooth Harrow	m	600	0.0	0.0	ft	41	0.0	0.0	1.0	1.0	1.0	30
Spring Tooth Harrow	m	2,000	0.0	0.0	ft	135	0.0	0.0	1.0	1.0	1.0	35
Roller Packer	m	600	0.0	0.0	ft	41	0.0	0.0	1.0	1.0	1.0	50
Roller Harrow	m	2,600	0.0	0.0	ft	180	0.0	0.0	1.0	1.0	1.0	50
Land Plane	m	8,000	0.0	0.0	ft	550	0.0	0.0	1.0	1.0	1.0	45
SEEDING IMPLEMENTS												
Row Crop Planter, prepared seedbed												
mounted												
seeding only	rows	500	0.0	0.0	rows	110	0.0	0.0	1.0	1.0	1.0	25
drawn												
seeding only	rows	900	0.0	0.0	rows	200	0.0	0.0	1.0	1.0	1.0	25
seed, fertilizer, herbicides	rows	1,550	0.0	0.0	rows	350	0.0	0.0	1.0	1.0	1.0	25
Row Crop Planter, no-till												
seed, fertilizer, herbicides												
1 fluted coulter/row	rows	1,820	0.0	0.0	rows	410	0.0	0.0	1.0	0.96	0.92	25
Row Crop Planter, zone-till												
seed, fertilizer, herbicides												
3 fluted coulters/row	rows	3,400	0.0	0.0	rows	765	0.0	0.0	1.0	0.94	0.82	35
Grain Drill w/press wheels												
< 2.4 m drill width	rows	400	0.0	0.0	rows	90	0.0	0.0	1.0	1.0	1.0	25
2.4 to 3.7 m drill width	rows	300	0.0	0.0	rows	67	0.0	0.0	1.0	1.0	1.0	25
> 3.7 m drill width	rows	200	0.0	0.0	rows	25	0.0	0.0	1.0	1.0	1.0	25
Grain Drill, no-till												
1 fluted coulter/row	rows	720	0.0	0.0	rows	160	0.0	0.0	1.0	0.92	0.79	35
Hoe Drill												
primary tillage	m	6,100	0.0	0.0	ft	420	0.0	0.0	1.0	1.0	1.0	50
secondary tillage	m	2,900	0.0	0.0	ft	200	0.0	0.0	1.0	1.0	1.0	50
Pneumatic Drill	m	3,700	0.0	0.0	ft	250	0.0	0.0	1.0	1.0	1.0	50

Draft force required to pull many seeding implements and minor tillage tools operated at shallow depths is primarily a function of the width of the implement and the speed at which it is pulled. For tillage tools operated at deeper depths, draft also depends upon soil texture, depth, and geometry of the tool. Typical draft requirements can be calculated as:

$$D = F*I [A + B + C(S)^2] W*T$$

where:

- ☆ D is implement draft, N (lbf);
- ☆ F is a dimensionless soil texture adjustment parameter (table below);
- ☆ i = 1 for fine, 2 for medium and 3 for coarse textured soils;
- ☆ A, B and C are machine specific parameters (table below);
- ☆ S is field speed, km/h (mile/h).
- ☆ W is machine width, m (ft) or number of rows or tools (table below);
- ☆ T is tillage depth, cm (in.) for major tools, or equals one for minor tillage tools and seeding implements.

Motion resistance becomes appreciable when heavy implements are used in soft or loose soils. Tire parameters and wheel loadings must be known or assumed to calculate this value. Total implement motion resistance is computed as the sum of the individual wheel values.

Drawbar power for tractor-powered implements (and propulsion power for self-propelled implements) is computed as:

$$Pdb = Ds / 3.6$$

where:

- ☆ Pdb is drawbar power required for the implement, kW;
- ☆ D is implement draft, kN;
- ☆ s is travel speed, km/h;

or,

$$Pdb = Ds / 375$$

where:

- ☆ Pdb is drawbar power required for the implement, hp;
- ☆ D is implement draft, lb;
- ☆ s is travel speed, mph.

Power-takeoff (PTO) power is power required by the implement from the PTO shaft of the tractor or engine. Typical PTO power requirements can be determined using rotary power requirement parameters available from a number of sources. Implement power take-off power can be calculated as

$$Ppto = a + bw + cF$$

where:

- ✰ Ppto is power-takeoff power required by the implement kW (hp);
- ✰ w is implement working width, m (ft);
- ✰ F is material feed rate, t/h(ton/h) wet basis;
- ✰ a, b, and c are machine specific parameters (Table below)

Table 2 - Rotary power requirement parameters

Table 2 – Rotary power requirement parameters

Machine Type	Parameter a kW	b kW/m	c kWh/t	Parameter a hp	b hp/ft	c hph/ton	Range[1] ±%
Baler, small rectangular	2.0	0	1.0[2]	2.7	0	1.2[2]	35
Baler, large rectangular bales	4.0	0	1.3	5.4	0	1.6	35
Baler, large round	4.0	0	1.1	5.4	0	1.3	50
Baler, large round	2.5	0	1.8	3.4	0	2.2	50
Beet harvester[3]	0	4.2	0	0	1.7	0	50
Beet topper	0	7.3	0	0	3.0	0	30
Combine, small grains	20.0	0	3.6[4]	26.8	0	4.4[4]	50
Combine, corn	35.0	0	1.6[4]	46.9	0	2.0[4]	30
Cotton picker	0	9.3	0	0	3.8	0	20
Cotton stripper	0	1.9	0	0	0.8	0	20
Feed mixer	0	0	2.3	0	0	2.8	50
Forage blower	0	0	0.9	0	0	1.1	20
Flail harvester, direct-cut	10.0	0	1.1	13.4	0	1.3	40
Forage harvester, corn silage	6.0	0	3.3[5]	8.0	0	4.0[5]	40
Forage harvester, wilted alfalfa	6.0	0	4.0[5]	8.0	0	4.9[5]	40
Forage harvester, direct-cut	6.0	0	5.7[5]	8.0	0	6.9[5]	40
Forage wagon	0	0	0.3	0	0	0.3	40
Grinder mixer	0	0	4.0	0	0	4.9	50
Manure spreader	0	0	0.2	0	0	0.3	50
Mower, cutterbar	0	1.2	0	0	0.5	0	25
Mower, disk	0	5.0	0	0	2.0	0	30
Mower, flail	0	10.0	0	0	4.1	0	40
Mower-conditioner, cutterbar	0	4.5	0	0	1.8	0	30
Mower-conditioner, disk	0	8.0	0	0	3.3	0	30
Potato harvester[3]	0	10.7	0	0	4.4	0	30
Potato windrower	0	5.1	0	0	2.1	0	30
Rake, side delivery	0	0.4	0	0	0.2	0	50
Rake, rotary	0	2.0	0	0	0.8	0	50
Tedder	0	1.5	0	0	0.6	0	50
Tub grinder, straw	5.0	0	8.4	6.7	0	10.2	50
Tub grinder, alfalfa hay	5.0	0	3.8	6.7	0	4.6	50
Windrower/swather, small grain	0	1.3	0	0	0.5	0	40

[1] Range in average power requirement due to differences in machine design, machine adjustment, and crop conditions.
[2] Increase by 20% for straw.
[3] Total power requirement must include a draft of 11.6 kN/m (±40%) for potato harvesters and 5.6 kN/m (±40%) for beet harvesters. A row spacing of 0.86 m for potatoes and 0.71 m for beets is assumed.
[4] Based upon material-other-than-grain, MOG, throughput for small grains and grain throughput for corn. For a PTO driven machine, reduce parameter a by 10 kW.
[5] Throughput is units of dry matter per hour with a 9 mm (0.35 in.) length of cut. At a specific throughput, a 50% reduction in the length of cut setting or the use of a recutter screen increases power 25%.

Hydraulic power is the fluid power required by the implement from the hydraulic system of the tractor or engine. Implements hydraulic power can be computed as

$P_{hyd} = pF/1000$

where:

- ☆ Phyd is hydraulic power required by the implement, kW;
- ☆ F is fluid flow, L/s;
- ☆ p is fluid pressure, kPa;

or,

$P_{hyd} = pF/1714$

where:

- ☆ Phyd is hydraulic power required by the implement, hp;
- ☆ F is fluid flow gal/min;
- ☆ p is fluid pressure, psi.

Electric power is required to operate components of some implements. To compute implement electric power

$P_{el} = IE/1000$

where:

- ☆ Pel is electric power required by the implement, kW;
- ☆ I is electric current, A;
- ☆ E is electric potential, V;

or,

$P_{el} = IE/746$

where:

- ☆ Pel is electric power required by the implement, hp;
- ☆ I is electric current, A;
- ☆ E is electric potential, V;

Total power requirement for operating implements (drawn or self propelled) is the sum of implement power components converted to equivalent PTO power. Total implement power requirement can be computed as

$PT = (P_{db}/ E_m E_t) + P_{pto} + P_{hyd} + P_{el}$

where:

- ☆ PT is total implement power requirement, kW (hp);
- ☆ Et is traction efficiency (expressed as a decimal);
- ☆ Pdb is drawbar power required for the implement, kW (hp);

- Phyd is hydraulic power required by the implement, kW (hp);
- Ppto is power-takeoff power required by the implement, kW (hp);
- Pel is electric power required by the implement, kW (hp);
- Em is mechanical efficiency of the transmission and power train.
- This coefficient is typically 0.96 for tractors with gear transmissions.

Additional power is required to accelerate and overcome changes in topography, soil and crop conditions. Additional power is also required for operator-related equipment such as hydraulic control systems, air conditioning, etc.

Field machine performance

Field efficiency is the ratio between the productivity of a machine under field conditions and the theoretical maximum productivity. Field efficiency accounts for failure to utilize the theoretical operating width of the machine; time lost because of operator capability and habits and operating policy; and field characteristics.

Travel to and from a field, major repairs, preventive maintenance, and daily service activities are not included in field time or field efficiency. Field efficiency is not a constant for a particular machine, but varies with the size and shape of the field, pattern of field operation, crop yield, moisture, and crop conditions. The following activities account for the majority of time lost in the field:

- turning and idle travel;
- materials handling;
  - seed;
  - fertilizer;
  - chemicals;
  - water;
  - harvested material;
- cleaning clogged equipment;
- machine adjustment;
- lubrication and refueling (besides daily service);
- waiting for other machines.

Effective field capacity is a function of field speed, machine working width, field efficiency, and unit yield of the field. Area capacity is expressed as

Ca = swEf/10

Table 3 - Field efficiency field speed and equal and parameters

Machine	Field efficiency Range %	Field efficiency Typical %	Field speed Range mph	Field speed Typical mph	Field speed Range km/h	Field speed Typical km/h	Estimated life h	Total life R&M cost % of list price
TRACTORS								
2 wheel drive & stationary							12 000	100
4 wheel drive & crawler							16 000	80
TILLAGE & PLANTING								
Moldboard plow	70–90	85	3.0–6.0	4.5	5.0–10.0	7.0	2 000	100
Heavy-duty disk	70–90	85	3.5–6.0	4.5	5.5–10.0	7.0	2 000	60
Tandem disk harrow	70–90	80	4.0–7.0	6.0	6.5–11.0	10.0	2 000	60
(Coulter) chisel plow	70–90	85	4.0–6.5	5.0	6.5–10.5	8.0	2 000	75
Field cultivator	70–90	85	5.0–8.0	7.0	8.0–13.0	11.0	2 000	70
Spring tooth harrow	70–90	85	5.0–8.0	7.0	8.0–13.0	11.0	2 000	70
Roller-packer	70–90	85	4.5–7.5	6.0	7.0–12.0	10.0	2 000	40
Mulcher-packer	70–90	80	4.0–7.0	5.0	6.5–11.0	8.0	2 000	40
Rotary hoe	70–85	80	8.0–14.0	12.0	13–22.5	19.0	2 000	60
Row crop cultivator	70–90	80	3.0–7.0	5.0	5.0–11.0	8.0	2 000	80
Rotary tiller	70–90	85	1.0–4.5	3.0	2.0–7.0	5.0	1 500	80
Row crop planter	60–75	65	4.0–7.0	5.5	6.5–11.0	9.0	1 500	75
Grain drill	55–80	70	4.0–7.0	5.0	6.5–11.0	8.0	1 500	75
HARVESTING								
Corn picker sheller	60–75	65	2.0–4.0	2.5	3.0–6.5	4.0	2 000	70
Combine	60–75	65	2.0–5.0	3.0	3.0–6.5	5.0	2 000	60
Combine (SP)[1]	65–80	70	2.0–5.0	3.0	3.0–6.5	5.0	3 000	40
Mower	75–85	80	3.0–6.0	5.0	5.0–10.0	8.0	2 000	150
Mower (rotary)	75–90	80	5.0–12.0	7.0	8.0–19.0	11.0	2 000	175
Mower-conditioner	75–85	80	3.0–6.0	5.0	5.0–10.0	8.0	2 500	80
Mower-conditioner (rotary)	75–90	80	5.0–12.0	7.0	8.0–19.0	11.0	2 500	100
Windrower (SP)	70–85	80	3.0–8.0	5.0	5.0–13.0	8.0	3 000	55
Side delivery rake	70–90	80	4.0–8.0	6.0	6.5–13.0	10.0	2 500	60
Rectangular baler	60–85	75	2.5–6.0	4.0	4.0–10.0	6.5	2 000	80
Large rectangular baler	70–90	80	4.0–8.0	5.0	6.5–13.0	8.0	3 000	75
Large round baler	55–75	65	3.0–8.0	5.0	5.0–13.0	8.0	1 500	90
Forage harvester	60–85	70	1.5–5.0	3.0	2.5–8.0	5.0	2 500	65
Forage harvester (SP)	60–85	70	1.5–6.0	3.5	2.5–10.0	5.5	4 000	50
Sugar beet harvester	50–70	60	4.0–6.0	5.0	6.5–10.0	8.0	1 500	100
Potato harvester	55–70	60	1.5–4.0	2.5	2.5–6.5	4.0	2 500	70
Cotton picker (SP)	60–75	70	2.0–4.0	3.0	3.0–6.0	4.5	3 000	80
MISCELLANEOUS								
Fertilizer spreader	60–80	70	5.0–10.0	7.0	8.0–16.0	11.0	1 200	80
Boom-type sprayer	50–80	65	3.0–7.0	6.5	5.0–11.5	10.5	1 500	70
Air-carrier sprayer	55–70	60	2.0–5.0	3.0	3.0–8.0	5.0	2 000	60
Bean puller-windrower	70–90	80	4.0–7.0	5.0	6.5–11.5	8.0	2 000	60
Beet topper/stalk chopper	70–90	80	4.0–7.0	5.0	6.5–11.5	8.0	1 200	35
Forage blower							1 500	45
Forage wagon							2 000	50
Wagon							3 000	80

[1]SP indicates self-propelled machine.

where:

- ☆ Ca is area capacity, ha/h;
- ☆ s is field speed, km/h;

- ☆ w is implement working width, m;
- ☆ Ef is field efficiency, decimal;

or,

Ca = swEf/8.25

where:

- ☆ Ca is area capacity, acre/h;
- ☆ s is field speed, mile/h;
- ☆ w is implement working width, ft;
- ☆ Ef is field efficiency, decimal.

Typical ranges of field efficiency and field speed can be found in Table 3.

Material capacity is expressed as

Cm = swyEf/10

where:

Cm is material capacity, t/h;

s is field speed, km/h;

w is implement working width, m;

Ef is field efficiency, decimal;

y is unit yield of the field, t/ha;

or,

Cm = swyEf/8.25

where:

Cm is material capacity, ton/acre;

s is field speed, mile/h;

w is implement working width, ft;

Ef is field efficiency decimal.

y is unit yield of the field, ton/acre.

Chapter 8

AGRICULTURE WORKING TOOLS

8.1 HARVESTING EQUIPMENT

8.1.1 Chopper

Fig Chopper

Chopper is a simple harvesting hand tool, which consists of a blade, tang and a handle. The cutting edge is slightly curved (convex) along the cutting edge to facilitate the cutting. The blade is joined to the handle by inserting its tang. The blade is forged to knife section and cutting edge sharpened. The blade is important part of the tool and is made from medium carbon steel, high carbon steel, manganese steel or alloy steel including tool steel. The cutting edge is hardened and tempered to suitable hardness to resist wear and abrasion. The tool made from tool steel and alloy steel has more service life as compared to the plain carbon steel blade. The blade should have property of toughness to resist the impact stresses. The handle is made of wood and shaped to suitable handgrip. For operation, the blade is struck hard against the stem with one hand and the cutting is accomplished by the impact action and sliding of the blade along the stem. The shape and size of the chopper varies from place to place and its use.

8.1.2 Tapioca harvester

The tapioca(cassava) harvesting machine features a multi-blade system for cutting and a conveyor system for transporting cut cassava plants to the area where they are laid out horizontally in rows. A hydraulic motor provides the source of power for the blade system and the conveyor system. The multi-blade system features 30 circular saws, with 20cm. in diameter each, arranged into top and bottom rows. The bottom blades are set slightly forward, in order to cut the trunk first, then the tip. Circular blades yield cleaner cuts and less damage to the plant. The PTO (power takeoff) is used to send power from the hydraulic motor to the blades and conveyor systems.

Initial testing using three speed levels (1,500, 2,000, and 2,500 cycles per minute) showed that the most efficient speed is 2,500 cycles per minute. At this speed, the tapioca(cassava) harvesting machine covered an area of 1 rai in 1 hour, with 22 percent damage to the crop, and consumed the fuel at 7 liters per hour. After the initial testing, the team made an improvement to the conveyor system, specifically the speed of the hydraulic motor, in order to slow down the transport function and achieve a more organized way of arranging the harvest onto the collector tray. A speed control valve was installed, to allow adjustment of the motor speed.

8.1.3 Tractor Mounted Vertical Conveyor Reaper

The machine consists of a 76 mm pitch reciprocating cutter bar assembly, seven crop row dividers, and two vertical conveyor belts fitted with lugs, pressure springs, pulleys and gearbox for the power transmission system. The crop row dividers are fitted in front of the cutter bar assembly and the star wheels are mounted over the crop row dividers. The machine is mounted in front of the tractor and the power to the machine is given from tractor PTO with the help of intermediate shaft running beneath the chassis of the tractor and a coupling shaft. Height of the machine above ground is controlled by tractor hydraulics with the help of pulleys and steel ropes. After the crop is cut by the cutter bar, it is held in a vertical position and crop is delivered to one side of the machine by lugged belt conveyors and fall on the ground in the form of a windrow perpendicular to the direction of movement of the machine.

8.1.4 Self-Propelled Walking Type Vertical Conveyor Reaper

It is an engine operated, walk behind type harvester suitable for harvesting and windrowing cereals like wheat and paddy and oilseed crops. The reaper consists of engine, power transmission box, lugged wheels, cutter bar, crop row dividers, conveyor belts with lugs, star wheels, operator's controls and a sturdy frame. The engine power is transmitted to cutter bar and conveyor belts through belt-pulleys. During forward motion of the reaper, crop row dividers divide the crop, which come in contact with cutter bar, where shearing of crop stems takes place. The cut crop is conveyed to one side of the machine by the conveyor belt fitted with lugs. The crop is bundled manually and transported to threshing yard. The field capacity of the machine is 0.15 - 0.17 ha/h. The cost of the machine is about Rs. 80,000. The cost of operation is Rs. 1150/ha with this machine as compared to Rs. 3200/ha by traditional method. There is a saving of 90-95% in labour and time and 63% in cost of operation as compared to conventional method.

8.1.5 Tractor Front Mounted Vertical Conveyor Reaper

The machine is used for harvesting cereal crops like wheat and paddy. It consists of a 76 mm pitch reciprocating cutter bar assembly, seven crop row dividers, and two vertical conveyor belts fitted with lugs, pressure springs, pulleys and gearbox for the power transmission system. The crop row dividers are fitted in front of the cutter bar assembly and the star wheels are mounted over the crop row dividers. The machine is mounted in front of the tractor and the power to the machine is given from tractor PTO. The height of the machine above ground is controlled by tractor hydraulic system with the help of pulleys and steel ropes. After the crop is cut by the cutter bar, it is held in a vertical position and delivered to one side of the machine by lugged belt conveyors and fall on the ground perpendicular to the direction of movement of machine in the form of a windrow. The field capacity of the machine is 0.4 ha/h at 2.5-3.5 km/h forward speed. The cost of operation is about Rs. 1600/- as against Rs. 3200/ha by traditional method.

8.1.6 Tractor Mounted Fodder Harvester

A tractor (41 kW) operated flail type harvester-cum-chopper-cum-loader was designed to facilitate harvesting, chopping and loading operations simultaneously. The crops viz. bajra, sorghum, maize, barseem and oats with height 1.00 to 2.50 m and stalk density 20-80 plants/m2 can be harvested successfully. The machine consists of a rotary shaft mounted with blades (flails) to harvest the crop, auger for conveying the cut crop, cutters for chopping and conveying chopped fodder through outlet into the trailer. The blades on the rotary shaft are staggered in three rows of 13 blades each on a horizontal axis perpendicular to the direction of motion. The blades cut the crop and auger conveys it to the chopper unit. The chopping mechanism having 4 blades cuts crops into pieces and the chopped material is thrown out at high speed and is filled into the hitched trailer to the machine. It also harvested lodged and over matured crop without any difficulty. The capacity of the machine is 0.20 ha/h at a forward speed of 2.5-4.0 km/h. The cost of operation of this machine is Rs. 1500/ha against Rs. 3200/ha by conventional method.

8.1.7 Tractor Mounted Root Crop Harvester cum Elevator

A root crop harvester-cum-elevator was developed by PAU, Ludhiana for digging onion and other root crops. The field capacity of the machine is 0.28, 0.24, 0.21 and 0.21 ha/h for digging of carrot, potato, garlic and onion crops, respectively at a forward speed of 2.78, 2.41, 2.10 and 2.10 km/h. The damage to tubers is 1.98, 1.92 1.22 and less than 1.0% for digging of carrot, potato, garlic and onion crops, respectively. The approximate cost of the machine is Rs. 60,000/-. The saving in cost of operation and labour for harvesting onion, carrot and garlic is 52.28, 46.71, 52.28%, and 69.05, 59.29 and 69.05%, respectively as compared to manual harvesting.

8.1.8 High Capacity Multi-crop Thresher

It is suitable for threshing wheat, maize, sorghum, rice, gram, pigeon pea, soybean, mustard, sunflower, safflower and linseed crops. It consists of spike tooth cylinder, aspirator type blower and sieve shaker. Two top covers, three concaves, three sieves and variable cylinder speeds (7-21 m/s) are provided for threshing different crops. Reciprocating sieve system consists of replaceable set of sieves and screens mounted on oscillating frame. The amplitude of oscillations could also be varied. Upper sieve separates chaff bigger than grain while lower screen let the dust and smaller particles than grain pass through thus help in delivering clean grain at main outlet. The thresher is provided with four wheels made of cast iron for transportation and motor stand to fit the motor on it. It also has provision for attaching universal shaft for operating the thresher by tractor PTO.

The grain output capacity is 16-20 q/h for wheat, 8-10 q/h for raya, 6-8 q/h for gram, 4-5 q/h for green gram, 4.0 q/h for moong and 5.0 q/h for guar crops. The machine is operated at the lowest cylinder speed of 300 rpm for moong to minimize the breakage while it is operated at 500 rpm for guar crop. Threshing efficiency, un-threshed grain and visible damage are 98-99%, 1.5-2.0% and 1.0%, respectively. Threshing efficiency and cleaning efficiency are observed to be within the recommended limit. An average total losses are about 5%. The cost of machine is Rs. 70,000/- to Rs. 80,000/ and operating cost is Rs. 250-300/h.

8.2 HARVESTING EQUIPMENT TROUBLESHOOTING, MAINTENANCE , SERVICING

It's always good to know the equipment and take advantage of the tips and techniques in the owner's manual or service manual that came with it. Good maintenance starts with inspecting machinery regularly, before and after harvest. There's no surprise that routine service includes checking and managing all lubricants and fluids, filters, cleaners and greasing. Plus, another good thing to remember is that it's always wise to keep equipment clean, including hot surfaces being kept free of oil and debris. For winter storage, clean equipment with air and avoid using water when it can get into spaces that can rust. A cloth wipe-down helps, too.

While you work to maintain equipment, please be sure to practice everyday safety tips that help prevent accidents. Routine fix-ups and safety reminders can seem obvious, but we truly appreciate our relationship and simply wish you a happy, healthy, injury-free work environment.

8.2.1 Troubleshooting in Harvesting Machines

Problem	Remedy
Cutterbar	It is absolutely necessary to stop the combine when working on or around the cutterbar. If it is necessary to work underneath, block the cutterbar upto avoid risk of accident.
Poor cutting action	1. Adjust knife registration. 2. Reset knife keeps. 3. Sharpen knife. 4. Straighten finger bar. 5. Replace damaged knife sections. 6. Remove build-up of foreign material from the cutterbar.
Sudden knife blockage	1. Remove foreign objects. 2. Replace misaligned cutterbar fingers. 3. Adjust tension of drive belt.

Problem	Remedy
Dirt and material built-up on divider points	1. Set drives to higher position using the slide. 2. Renew cutter bar skids if necessary. 3. Adjust the reel further forward.
Cutterbar lifts too slowly	1. Check hydraulic oil level and fill up if necessary. 2. Have hydraulic oil pressure checked.
Reel tending to stop	Adjust slip clutch tension a little bit.
Feed rake and cutterbar tending to stop	Adjust jockey pulley of cutter bar V-belts clutch.
Cutterbar dips at one end	Loosen the securing bolts on the feeder house and adjust cutterbar mountings to level cutterbar.
Cutterbar clutch will not disengage	1. Check adjustment of cutterbar clutch belt guides. 2. Clean V-belt pulleys of cutterbar clutch.
Uneven crop feed	1. Adjust main table auger height to suit crop conditions. 2. Adjust reel tine position and reel speed. 3. Adjust feed rake chains to proper tension.
Main table auger tending to stop or gets blocked	Turn the main table auger back by the flights and remove foreign matter.
Crop wrapping round the reel shaft ends.	Adjust inner deflectors closer to cutterbar centre and possibly higher.
Crop wrapping round the reel finger tubes.	1. Raise the reel. 2. Set reel fingers more inclined to the front. 3. Adjust reel speed to match ground speed.
Reel fingers fouling knife in lowest position	Raise reel position by means of reel cylinder eye bolts.
Reeldipsatoneend	Check for equal adjustment of reel cylinder's eye bolts.
Too many stones being picked-up	1. Reduce number of grain lifters, if fitted. 2. Do not cut unnecessarily low (lodged crops should be lifted with grain lifters).

Problem	Remedy
Slugging of drum	1. Reduce drum-to-concave clearance 2. Increase drum speed in damp crops 3. Adjust basic settings of concave 4. Care for even feeding of crop to feed rake 5. Adjust main table auger further down 6. Check and have speed of main drive shaft adjusted 1140±5 rpm if necessary 7. Reduce ground speed as moisture level increases 8. Adjust spring-loaded cylinder for threshing mechanism drive power band belt stones and other foreign matter causing damage to threshing parts
Stones and other foreign matter causing damage to threshing parts	1. Clear stone trap frequently 2. Do not cut unnecessarily low on stony ground
Irregular speed variations	1. Correct the settings of the idler pulley and spring-loaded cylinder that controls the tension of the power band belt 2. Check engine (see engine problems)
Poor threshing action	1. Adjust concave closer to drum 2. Increase drum speed 3. Adjust basic setting of concave 4. Have worn or damaged threshing equipment repaired or replaced. 5. Adjust relative position of concave entrance and concave exit to match crop conditions 6. Use special threshing equipment if necessary.
Unbalance	1. Clean dirty threshing drum 2. Remove dust accumulations from belt pulleys 3. Have damaged rotor repaired
Grain cracking	1. Decrease drum speed 2. Increase drum-to-concave clearance 3. Adjust frogmouth sieves wider open 4. Adjust tension of elevator chains
Excessive belt wear	1. Engage combine drives with engine at slow idling speed and then bring the engine to full speed 2. Adjust belt tensioners

8.2.2 Maintenance and Servicing

1 Moving Equipment

- ☆ All equipment should be washed thoroughly before it is moved to a new operation area to prevent the spread of weeds, pests and disease.
- ☆ Care must be taken to avoid spillage of fuel or oil during transportation.

2 Workshop Facilities

- ☆ Locate facilities at least 50 m from any water body or watercourse.
- ☆ Drainage from workshop areas should be diverted into collecting/settling ponds.
- ☆ Non-toxic solid waste should be removed or buried and covered with at least 0.5 m of soil.

3 Toxic Waste

- ☆ The use of toxic substances is to be minimised and wherever possible biodegradable substitutes used.
- ☆ Toxic materials include spent hydraulic and refrigeration fluid, oil, fuel (petrol, diesel), workshop wastes, oily water separator sludge and residues, industrial cleaners, paints and resins, preservatives (including timber treatment chemicals), distillates, lubricants, insecticides, and biocides.
- ☆ Toxic materials are to be collected in containers securely sealed and preferably returned to municipal/township disposal areas, returned to the manufacturer or buried in an appropriate waste pit.
- ☆ Where onsite burial of toxic materials and their containers is the only alternative, the location of the pit should be clearly indicated (including signs in all the local languages), and the pit securely fenced.
- ☆ All waste pits are to be covered with at least 0.5 m of soil and located at least 100 m from maximum inundation by water and at least 1 m above maximum groundwater height.

4 Main Fuel and Oil Storage

- ☆ Locate main fuel and oil storage in a well drained area at least 50 m from any watercourse; and no closer than 100 m to village areas.
- ☆ Bunds (preferably concrete mounds) with a capacity of twice (× 2) the storage capacity must be provided around the storage.
- ☆ Drains are to be directed to a closed, stable and flood free disposal pit, at least 50 m from a watercourse or water body.

- ☆ Where main fuel storage facilities must be located on a wharf, they should be located as far away from the sea as the fuel pumping lines permit and surrounded by a leak-proof bund, which in the event of a spill can contain all fuel so that it can later be pumped to a disposal pit at least 50 m from a water body.

5 Field Servicing and Maintenance

- ☆ Field fuel tanks, refuelling points and maintenance areas are to be located in well-drained areas, such as landings or road junctions; outside areas excluded from harvesting; more than 50 m from a watercourse.
- ☆ Care must be taken to prevent spillage during refuelling.
- ☆ Sump oil should not be dumped in the harvesting areas. It should be collected and removed to the main disposal facility.

6 Refuse Disposal

- ☆ All solid waste should be placed in a refuse pit and buried as soon as harvesting in the coupe is completed.

Figure: Removal of all Refuse from Harvesting Areas

Things to be checked in harvester equipments before using:

1. Clean the diesel tank and top up with clean diesel
2. Change all diesel filter elements.
3. Change the engine oil and oil filter.
4. Check and reset tappet clearance.
5. Check the injection pressure and spray of the injectors.
6. Change air cleaner oil, clean air cleaner bowl and refill it with recommended grade of oil.
7. Change hydraulic oil and top-up with the recommended grade of oil.
8. Change sieves and straw walker.
9. Check battery for its voltage, electrolyte level and gravity. If required add only distilled water to make up the electrolyte level.

10. Lubricate all greasing points.
11. Check oil in transmission and hydraulic system.
12. Check flat belts, V-belts and roller chains for proper tension.
13. Check water level in the radiator. Use only fresh and clean water.
14. Check tyre pressure and reset the tyre inflation (front tyre: 0.18 to 0.19 MPa and rear tyre0.25 to 0.28 MPa)
15. Check~ all nuts and bolts for their proper torque and locking.
16. Check clearance between feeder bottom and feeder angle.
17. Check clearance between 'cutter bar pan and conveyor worm.
18. Check the clearance between threshing drum and concave at front and rear.
19. Check all electrical components for their proper functioning.
20. Start the engine and check;

    (a) leakages' from diesel, hydraulic, engine oil connections and

    (b) Engine oil pressure, battery charging, and rate system running.
21. Check all ball and taper roller, thrust and needle roller bearings.
22. Check the chain for broken links.
23. Check welding joints and broken components for replacement. Check an shaft for any crack, bend or undersize/loose bearing surface.

General maintenance:

Brake:

A. path of brake pedal too long due to worn out or not properly adjusted brake linings	1. Replace the lining or adjust the brakes
B. Brake pedal does not return effectively because tension spring has lost its tension or elongated.	1. Replace the tension spring
C. Insufficient breaking action due to 1. Brake lining contaminated with grease or oil 2. Water may have penetrated in the brake housing 3. Brake assembly bolts are loose.	 1. Replace the brake lining 2. To vaporize the water apply brakes several times 3. Tighten the bolts
D. Pulling towards one side due to brake adjustment of one side, either too tight or too loose. One side lining contaminated with grease or oil	1. Readjust the brakes equally

Hydraulic system

A hydraulic cylinder lifts the cutter bar too slow due to

Problem	Remedy
1. Hydraulic pump belt loose	1. Set to proper tension
2. Improper grade of oil	2. Use proper grade oil
3. Leakage in the system	3. Set it right
4. Blockage in the system	4. Correct it
5. Air in the system	5. Bleed it

A. Premature belt wearing / belt flickering/ jockey pulley support breaking due to

Problem	Remedy
1. Wrong alignment	1. Set right alignment
2. Fouling of belt	2. Correct fouling
3. Improper spring tension	3. Set the specified tension
4. Wrong pulleys fitted	4. Put right pulleys

B. Premature wearing of chains and sprockets' due to

Problem	Remedy
1. Alignment not OK	1. Set it right
2. Chain loose	2. Tighten the chain

8.2.3 Cleaning

Problems	Remedy
Uneven feed to sieves	Clean preparation floor
Knocking sieve pan	1. Remove dirt accumulations from above cleaning fan housing 2. Retighten all bolts that secure the sieve pan 3. Tighten the components securing the sieves 4. Have sieve pan bearings replaced
Excessive built up of material on frogmouth sieves	1. Reduce drum speed 2. Adjust concave to drum clearance wider open 3. Open concave clearance at the rear 4. Adjust frog mouth sieves to somewhat closer gaps 5. Increase wind blast

Problems	Remedy
Dirty grain sample	1. Increase wind blast
	2. Set air Intake apertures to correct position
	3. Decrease frog mouth sieves opening.
	4. Decrease main frogmouth sieve opening, respectively
	5. Have basic speed of machine checked and corrected I~ necessary (belt tension).

8.3 TILLAGE EQUIPMENTS

8.3.1 Introduction

- Tillage is the preparation of the soil for planting and the process of keeping it loose and free from weeds during the growth of crops.
- The primary objectives of fundamental purposes of tillage are divided into three phases: (1) to prepare a suitable seedbed, (2) to destroy competitive weeds and (3) to improve the physical condition
- of the soil.
- The basic tillage operation is the breaking of the soil in the preparation of a suitable seedbed.
- The breaking and loosening of the soil is the oldest phase of village as it includes the various types of plows.
- Prehistorically, it is assumed that man used crude tools of wood or other material with which he could loosen the soil.
- Perhaps a broken branch of a tree was the first tillage tool available.
- Later he learned to use fire or stone hand tools to fashion a soil-strirring tool from a fork of a tree by burning or hacking off one branch, leaving the longer for the beam and the trunk part for a handle.
- Still later, he was able to use miral power to pull the plow.

8.3.2 Background

- The plow is probably the oldest agricultural tool.
- The recorded history in the form of hieroglyphs and cuneiform characters shows that the ancients had a plow thousands of years B.C.
- It is recorded that about 900 B.C.
- Elisha as found “plowing with twelve yoke of oxen before him”.

- ☆ The wooden plow with a metal share has been used for many centuries, and millions of wooden plows are still is use today.
- ☆ The parts of the wooden plow were usually held together by animal thongs as there were no nails, bolts, or haywire available.
- ☆ The Roman plow which was improved by the Dutch, was imported into England about 1730.
- ☆ The Ensex plow of about 1756 had an iron moldboard.
- ☆ The Norfolk sheel plow of 1721 had a cast-iron share and an iron rounded moldboard.
- ☆ A curved moldboard made its appearance in 1760 on the Suffolk swing plow.
- ☆ The Rotherham plow was improved by James Small, who wrote a book on plow design in 1984.
- ☆ The close of the eighteenth century saw the chage in England from the wooden plow to the iron plow.
- ☆ In American, Thomas Jefferson and Daniel Webster were among the first to advance improvements of the plow.
- ☆ Charles Newbold of Burlington, New Jersey, secured the first patont on a cast-iron plow in 1797.
- ☆ Farmers rejected this iron plow because say thought it poisoned the soil; Jethro Wood developed a moldboard in 1814 of such curvature as to turn the soil in even furrows.
- ☆ The first steel plow was made from three sections of an old handsaw by John Lane about 1833.
- ☆ He also secured in 1868 a patent for soft-center steel, which is used at the present time in making moldboard for plows. In 1837, John Deere at Grand Detour, Illinois, made a steel plow (share and moldboard in one piece) from an old sawmill saw.
- ☆ Ten years later be established factory at Moline, Illinois.
- ☆ James Oliver was granted a patent in 1868 for hardening cast iron which was own as chilled iron.
- ☆ In 1856, M. Furley patented a single-bottom sulky or wheel plow which permitted the operator to ride.
- ☆ In 1864, F.S. Davenport patented a riding two-bottom horse drawn gang plow.

- ☆ Three-and four-bottom gang plows often required ten to twelve horses to pull them.
- ☆ The large ten-to fifteen-bottom plows were pulled by steam tractors in the 1890's by the large, slow, cumbersom gasoline-engine tractors from about 1900 to 1910.
- ☆ The early two-to five-bottom trailing-tractor plows were equipped with hand-lever lifts.
- ☆ In the early twenties, mechanical power lifts were developed.
- ☆ They were used until the hydraulic lift was introduced in the forties.
- ☆ The integral tractor-mounted unit assembly and unit-lifted plow were developed in the early forties by Ferguson.
- ☆ This type of plow is now becoming popular on small and average sized farms.

Disc plow

- ☆ Was probably developed about 1890. Models were listed in implement catalogues by 1895.
- ☆ One of the earliest patents for a disk plow was secured by M.A. and I.M. Cravath, Bloomington, Illionois. J.K. Underwood, D.H. Lane, and M.T. Hancock made improvements on the disk plow and made it practical.
- ☆ Since 1900, the development of the disk plow has followed trends similar to that of the moldboard plow.

Influence of the flow on man

- ☆ When man grasped a crooked stick and began to till the soil, he took his first step toward civilization.
- ☆ With each phase in the development of the plow, there has been a corresponding advance in civilization.
- ☆ In the beginning, one man, even though he gave all his time and energy to the task, could till only a small acreage.
- ☆ Later, animal power was applied and the acreage per man was increased.
- ☆ Now, with the large amount of mechanical power available, the acreage per man has been very materially increased.
- ☆ Thus, man can now produce more foods tuff than is necessary for his own sustenance and can furnish food to many who are working at other tasks.
- ☆ Hence, we can say that the flow is the foundation of civilization.

- ☆ In the production of all kinds of crops and in the preparation of a seedbed for them, the plow is the first tool used, and it is thus the basic tool of the farm.
- ☆ With the plow the ground is broken and pulverized into small particles, trash on the surface may be left on the surface or completely covered.
- ☆ One not familiar with the nature of the soil, the influence of water, air, and temperature upon its physical condition, and the action of the plow upon it may think that the plow is a very simple tool, requiring very little adjustment and practically no care at all.
- ☆ But those who are familiar with soil conditions and the plow adjustments necessary to obtain the best results know that good planning is not a simple matter.
- ☆ Optimum plowing requires the consideration of numerous factors.

8.3.3 Types of tillers

- ☆ Disc tillers vary considerably in the flexibility of the disc cylinder.
- ☆ The rigid type has a single axle.
- ☆ Some larger tillers have several independent disc cylinders or gangs.
- ☆ To provide flexibility these gangs all operate at the same angle, but each gang can move up and down separately over uneven terrain.
- ☆ Another flexible type has the axle jointed at intervals, making a snake-like flexible cylinder for working uneven ground.
- ☆ The flexible types can be shortened to suit the power available.
- ☆ The trailing type is pulled from the tractor drawbar.
- ☆ It is supported entirely by its won three wheels
- ☆ The semi-mounted tiller has a two-point, fairly rigid attachment to the tractor drawbar.
- ☆ The load is partly on the drawbar and partly on the rear furrow wheel

8.4 KINDS OF TILLAGE EQUIPMENT

- ☆ Tillage equipment can be divided into two general classes, namely, (1) primary tillage equipment and (2) secondary tillage equipment.
- ☆ Equipment that is used to break deeply and loosen the soil to prepare a suitable seedbed may be considered as primary tillage equipment. This group includes various kinds of moldboard ploughs, disc ploughs, subsoil ploughs or chisles, disc tiller ploughs and rotary tillers.

- ☆ Secondary tillage equipment includes harrows, pulverizers, cultivators, weeders, and special tools for surface tillage to conserve moisture and destroy weeds.

8.4.1 Secondary tillage

The term secondary tillage in this discussion means stirring the soil at a comparatively shallow depth (0 to 6 inches).

- ☆ The action of the plough does not usually prepare an ideal seedbed.
- ☆ Unless the plough is working in an ideal, mellow soil, further pulverization of the soil will generally be necessary before planting.
- ☆ As the young plants develop certain amounts of cultivation and weed eradication will be required.
- ☆ The equipment used to perform these operations are known as secondary tillage implements.
- ☆ The first implements to prepare the soil for sowing probably originated simultaneously with primitive ploughs.
- ☆ They were dry branches of various shapes drawn by men or occasionally by animals.
- ☆ It was not until the Roman era that the toothed harrow with wooden teeth (Fig. 151) appeared.
- ☆ In time, the shape of the harrow frame was altered.
- ☆ In the Middle Ages, harrows consisting of several bars tied crosswise were used. at the beginning of the nineteenth century, wooden harrows with rigid tines were produced on the pattern of harrows with spring tines.
- ☆ At the end of the nineteenth century cultivators for deep tillage were designed and were used to replace ploughs to some extent.
- ☆ About then the Higganum Corporation at Higganum, Connecticut, began making disc harrows commercially.
- ☆ The disc blades used before 1886 were all plain, perfectly smooth around the edge.
- ☆ In that year, George N. Clark, president of the Higganum Corporation, cut curved ntoches in the rim of the discs and used the term “cutway” to describe this disc harrows.
- ☆ Regular disc harrows, had become quite popular by 1900.
- ☆ In 1925, the offset disc harrow was designed to work under the branches of fruit trees.

Objectives

- ☆ To improve the seedbed by greater pulverization of the soil.
- ☆ To conserve moisture by summer-fallow operations to kill weeds and reduce evaporation.
- ☆ To cut up crop residue and cover crops and mix vegetable matter with the top soil.
- ☆ To break up clods, firm the top soil and put it in better tilth for seeding and germination of seeds.
- ☆ To destroy weeds on fallow lands.
- ☆ In general a good secondary tillage operation is possible after proper adjustment. An even better job is dependent upon the attention shown to the operation by the operator. The pictures below give an idea of a desirable soil consistency after harrowing.

Secondary tillage equipments

Types of secondary tillage equipment

- ☆ There are many types of machines that can be used for secondary tillage.
- ☆ They are the various types of harrows (disc, spike tooth, spring tooth), cultivators, land rollers and packers and rotary hoes.

Types of disc harrows

- ☆ The disc harrow first attained wide popularity during the latter part of the nineteenth century.
- ☆ It is now second only to moulboard plough in its importance as a tillage implement in dry land farming.
- ☆ Disc harrows may be classified according to their gang configuration as in Fig. 153 or according to whether they are trailed or mounted.

8.5 TYPES OF TILLAGES

There are various types of tillage.

Minimum Tillage	It is the minimum soil manipulation necessary to meet tillage requirements for crop production
Strip Tillage	It is a tillage system in which only isolated bands of soil are tilled.
Rotary Tillage	It is the tillage operations employing rotary action to cut, break and mix the soil.

Mulch Tillage	It is Preparation of soil in such a way that plant residues or other mulching materials are specially left on a near the surface.
Combined Tillage	Operations simultaneously utilizing two or more different types of tillage tools or implements to simplify, control or reduce the number of operations over a field are called combined tillage. Tillage is performed by tool, implement or machine.
Tool	It is an individual working element such as disc or shovel.
Implement	It is equipment generally having no driven moving parts, such as harrow or having only simple mechanism such as plough.
Machine	It is a combination of rigid or resistant bodies having definite motions and capable of performing useful work.

8.5.1 Minimum tillage equipment

- ☆ The following paragraphs suggest ways in which minimum tillage can be accomplished.
- ☆ Primary and secondary tillage, as well as planting, are operations that an be accomplished simultaneously under certain conditions.
- ☆ It is economically convenient and less soil compaction occurs, but some other factors may indicate the convenience of performing these operations at different times.

Plough Packer

- ☆ This minimum tillage operation uses a roller packer behind the mouldboard plough, Both implements are operated by the same tractor.
- ☆ The plough-packer works best on loam and lighter soils having some cover other than heavy sod.
- ☆ The plough packer does as little tillage as any of the once-over implements.
- ☆ Experiments, however, have shown that it is sufficient as long as adequate crop stands are obtained.
- ☆ This tiller is especially good for heavy soils.
- ☆ The blades press the furrows down and exert some packing below the surface.
- ☆ At the same time, the tune does some smoothing.
- ☆ This action does more to correct ploughing defects than the plough packer.
- ☆ Draft is about twice that needed for the packer.
- ☆ The packer can also be used in front of the planter during the next operation, but it is not so effective because the clods are drier and harder to break.

- ☆ A spite harrow can also be used instead of the packer.
- ☆ It can do a good if its width is twice that of the plough so as to cover the soil twice.
- ☆ The spike harrow is known by all farmers, and most of them have one probably discarded.
- ☆ One section makes a good light-draft minimum tillage implement.
- ☆ The teeth must be set slanting so the frame drags and levels the soil.
- ☆ It is convenient to hitch the harrow in such a way that it is lifted together with the plough.
- ☆ It is important because spike harrows collect trash.
- ☆ The rotary hoe, run backward, serves very wall as a minimum tillage tool.
- ☆ It crumbles lumps and presses down partly turned furrows.
- ☆ The rotary hoe, run backward, does a little presenting of furrows and breaks soft lumps.
- ☆ In amount of tillage performed and in draft, it compares favorably with the plough packer.

Circuit tiller

- ☆ Ttills and firms the soil by the turning action of a wheel which lies almost flat on the ground.
- ☆ The slight pitch of the wheel causes it to turn.
- ☆ A small gauage wheel holds it in position.
- ☆ Figure 39.
- ☆ This circular tiller revolves as it is pulled behind the plough.
- ☆ You can add weight to the wheel for more firming.

Spring tooth harrow

- ☆ it does not do a satisfactory job because it pulls up trash and grass.
- ☆ On the other hand, the draft is considerably higher than for any of the tools described above.

Plough packer with grain drill

- ☆ A grain drill covering just the width of the plough is hitched behind the plough packer unit described before.
- ☆ It works well when ploughing can be done at planting times. In some cases the soil may be too dry at that time for ploughing.

- Heavy compacted soils may have to be ploughed some time before planting in order to store some water for germination and the first stage of plant growth.

Rotary tillers development of history

- Although there has been a marked increase of interest in rotary plow and tillers in the United States during the last few years, the principle of rotary tillage is not new.
- Patents on devices of this type were issued as early as 1850-1860, but it was not until about 1910 that a successful rotary tiller using the elastic-claw principle with a staggered arrangement, was developed by Von Meyenberg of Switzerland.
- Swissrade rotary plows or tillers were introduced into the United States in about 1930, and soon afterwards several American manufacturers started making this type of equipment.
- It has only been rather recently that this type of tillage has experienced any real interest.
- The reason for the lack of interest was high cost, high power requirements and poorer weed control.

8.5.2 Types of rotary tillers

Fig Rotary tillers

Garden type

- self-propelled units (two wheel, 8 to 30 inches wide and 6 to 15 H.P. engines).
- The garden-type rotary tillers have proved themselves practical and effective for preparing seedbeds from unploughed ground for nurseries, greenhouses, vegetable crops, lawn planting, and similar jobs.
- These narrow machines are also suitable for shallow cultivation and weed control between rows of certain crops.

Trailed or tractor-mounted units with p.t.o. drives

- This type is usually 3 to 4 feet wide and requires 10 to 15 horese power for each foot of width.
- The cutting knives or tines are generally mounted on a horizontal power driven shaft which operates at about 300 r.p.m.
- The knives on some machines are provided with a shock-cushioned friction clutch that prevents the knives from breaking when they come in contact with a rock o solid obstacle.

Trailed units with auxillary engines

- This rotary tiller is pulled forward by a tractor but has the cutting knives driven by an auxillary engine mounted on the frame of the plough.
- This type of plough is made in 4 to 7-foot sizes and requires 60 to 100 horsepower.
- The cutting knives are mounted on a horizontal power-driven shaft.

Field-type, self-propelled machines (0 to 8 ft. cut, with engines as large as 180 h.p.

- Rotary tillers may also be classified by the position of the soil engaging rotor.
- Most of the rotary tiller have a horizontal axis for the rotating knives
- In Europe, tillers with a vertical axis for the rotating knives are also common
- There are some differences in the designers of frame work and drive mechanism to the two types.
- There are also some slight differences in their particular use of operation, but basically they do the same job and have the same limitations.

Development

- The disc tiller is also called the disc-plough, vertical disc-plough, one-way disc-plough, harrow plough, Wheatland plough, cylinder plough, tiller disc-plough and one-way disc. i.e. consider "disc tiller" the most suitable name.
- The disc tiller was developed in the Great Plains area. It was designed primarily as a one-way disc harrow. As its use spread, farmers began to adopt it for shallow ploughing. Improvements have made the tool into a popular and widely used plough.
- The speed of the plough should not be over 4 m.p.h. High-speed operation increases the power requirements, causes too much pulverizing of the surface soil, and does not leave trash on the surface to prevent wind erosion.

- ☆ This implement is about halfway between the standard disc-plough and the disc harrow in its soil working actions.
- ☆ It carries the discs at approximately a 45-degree angle to the travel direction cutting the throwing all the soil in the same direction. The discs are smaller than on a disc-plough, usually 16 to 24 inches in diameters, and have no scrapers to assist in inverting the soil.
- ☆ They cut shallower and narrower than those on the disc-plough, usually making a cut approximately 4-inches in width. They mix the soil with the trash instead of inverting the furrow slice and covering the trash.
- ☆ This mixing action retards drifting and blowing and keeps the soil in excellent condition to receive and hold moisture. Because the disc tiller is primarily a mixing tool, thc draft is lower per foot than with the disc-plough.
- ☆ Thus the same power, can pull a wider implement, cover more ground, and work faster than with the disc-plough. Where trash mixed through the top soil is not objectionable and where it is not necessary to work the soil deeply, a disc tiller can be substituted for a disc or moldboard plough.
- ☆ As the name Wheatland suggests, these implements were originally developed for the preparation of wheat land. The disc tiller is not a universal tool like the moldboard plough; it is limited to certain regions.

8.5.3 Disc plough

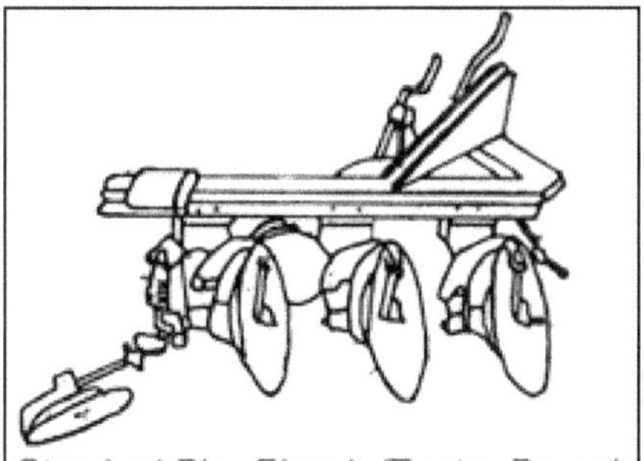

Standard Disc Plough (Tractor Drawn)

- ☆ It is a plough which cuts, turns and in some cases breaks furrow slices by means of separately mounted large steel discs.
- ☆ A disc plough is designed with a view to reduce friction by making a rolling plough bottom instead of sliding plough bottom.
- ☆ A disc plough works well in the conditions where mould board plough does not work satisfactorily.

Advantages of disc plough

1. A disc plough can be forced to penetrate into the soil, which is too hard and dry for working with a mould board plough.
2. It works well in sticky soil in which a mouldboard plough does not occur.
3. It is more useful for deep ploughing.
4. It can be used safely in stony and stumpy soil without much danger of breakage.
5. A disc plough works well even after a considerable part of the disc is worn off in abrasive soil.
6. It works in loose soil also (such as peat) without much clogging.

Disadvantages of disc plough

1. It is not suitable for covering surface trash and weeds as effectively as mouldboard plough does.
2. Comparatively, the disc plough leaves the soil in rough and more cloudy condition than that of smouldboard plough.
3. Disc plough is much heavier than mouldboard plough for equal capacities because penetration of this plough is affected largely by its weight rather than suction.
4. There is one significant difference between mouldboard plough and disc plough i.e. mould board plough is forced into the ground by the suction of the plough, while the disc plough is forced into the ground by its own weight.

Types of disc plough

Disc ploughs are of two types:

(i) Standard disc plough and (ii) Vertical disc plough.

Standard disc plough

- ☆ It consists of steel disc of 60 to 90 cm diameter, set at a certain angle to the direction of travel. Each disc revolves on a stub axle in a thrust bearing, carried at the lower end of a strong stand which is bolted to the plough beam (Fig. 8).
- ☆ The angle of the disc to the vertical and to the furrow wall is adjustable.
- ☆ It action, the disc cuts the soil, breaks it and pushes it sideways.
- ☆ There is little inversion of furrow slice as well as little burying of weeds and trashes.

- The disc plough may be mounted type or trailed type. In mounted disc plough, the wheels of the tractor take the side thrust.
- Sometimes a rear wheel is fitted to take side thrust of the plough to some extent.
- It trailed type; the furrow wheel of the plough takes side thrust.
- Disc is made of heat-treated steel of 5 mm to 10 mm thickness.
- The edge of the disc is well sharpened to cut the soil.
- The mount of concavity varies with the diameter the disc.
- The approximate values being 8 cm for 60 cm diameter disc and 16 cm for 95 cm diameter.
- A few important terms connected with disc plough explained below:

Disc

- It is a circular, concave revolving steel plate used for cutting and inverting the soil.

The rotovator

- It consists of 16 curved tines bolted on a horizontal power
- driven shaft.
- On this shaft 8 times are bent at edge in right hand side and another eight in the left hand side.
- The times are so arranged to give uniform cutting load on the shaft.
- Rotovator ploughs, pulverises mixes and levels dry and wet lands.
- Rotovator is used for pudding also in the previously ploughed land.
- It replaces cultivator, disc harrow and puddler.
- It converts crop residuals as manure.
- Rotovator offers a package of benefits to the farmers like getting a better seed bed, weed control, improvement of humus content and soil structure.

Disc tiller development

- The disc tiller is also called the disc-plough, vertical disc-plough, one-way disc-plough, harrow plough, Wheatland plough, cylinder plough, tiller disc-plough and one-way disc. i.e. consider “disc tiller” the most suitable name.
- The disc tiller was developed in Great Plains area.

- ☆ It was designed primarily as a one-way disc harrow.
- ☆ As its use spread, farmers began to adopt it for shallow ploughing.
- ☆ Improvements have made the tool into a popular and widely used plough. The speed of the plough should not be over 4 m.p.h.
- ☆ High-speed operation increases the power requirements, causes too much pulverizing of the surface soil, and does not leave trash on the surface to prevent wind erosion.
- ☆ This implement is about halfway between the standard disc-plough and the disc harrow in its soil working actions.
- ☆ It carries the discs at approximately a 45-degree angle to the travel direction cutting the throwing all the soil in the same direction.
- ☆ The discs are smaller than on a disc-plough, usually 16 to 24 inches in diameters, and have no scrapers to assist in inverting the soil.
- ☆ They cut shallower and narrower than those on the disc-plough, usually making a cut approximately 4-inches in width.
- ☆ They mix the soil with the trash instead of inverting the furrow slice and covering the trash.
- ☆ This mixing action retards drifting and blowing and keeps the soil in excellent condition to receive and hold moisture.
- ☆ Because the disc tiller is primarily a mixing tool, the draft is lower per foot than with the disc-plough. thus the same power, can pull a wider implement, cover more ground, and work faster than with the disc-plough. Where trash mixed through the top soil is not objectionable and where it is not necessary to work the soil deeply, a disc tiller can be substituted for a disc or moldboard plough.
- ☆ As the name Wheatland suggests, these implements were originally developed for the preparation of wheat land.
- ☆ The disc tiller is not a universal tool like the moldboard plough; it is limited to certain regions.

Types of disc ploughs

- ☆ As with moldboard ploughs, disc ploughs are also classified by (1) their method of hitching to the tractor (i.e., direct mounted, semi-mounted and trailing) and (2) whether to plough is reversible or not.

Integral mounted ploughs

- ☆ They are usually to the tractor by three-point linkages.
- ☆ They can be raised or lowered by the hydraulic system.
- ☆ Most can be attached in less than a minute.
- ☆ They are usually rear-mounted and have a rear wheel to absorb the side thrust.
- ☆ But some are mounted ahead of the rear wheels.
- ☆ A rear furrow wheel on an integral mounted disc plough may serve to counteract side pressures, hold the plough in alignment, and act as a gage wheel for ploughing depth.
- ☆ The depth in some makes is controlled by adjusting the hydraulic lift.

Semi-mounted disc plough

- ☆ The semi-mounted disc plough is also called a direct-connected blow.
- ☆ The front of this plough is connected to and mounted on the tractor, thus diminating the front furrow wheel and the land wheel.
- ☆ A furrow steel supports the rear end.
- ☆ This close-coupled plough is compact and easy to handle.
- ☆ It is easy to maneuver because short turns can be made, enabling the operator to close to fences.
- ☆ It can also be backed into corners.
- ☆ The rear wheel is automatically controlled from the steering mechanism at the front of the plough hydraulic lift raises the front of the plough high enough so it can be turned and transported easily.

Development History

- ☆ Although there has been a marked increase of interest in rotary plough and tillers in the United States during the last few years, the principle of rotary tillage is not new.
- ☆ Patents on devices of this type were issued as early as 1850-1860, but it was not until about 1910 that a successful rotary tiller using the elastic-claw principle with a staggered arrangement, was developed by Von Meyenberg of Switzerland.

- Swissrade rotary ploughs or tillers were introduced into the United States in about 1930, and soon afterwards several American manufacturers started making this type of equipment
- It has only been rather recently that this type of tillage has experienced any real interest. The reason for the lack of interest was high cost, high power requirements and poorer weed control.

8.6 USE AND MAINTENANCE OF TILLAGE EQUIPMENTS

8.6.1 Introduction

- The tillage tools remove the weeds by cutting.
- These tools are covered, dragged out from the surface of soil and are finally killed.
- Tillage tools loosen the soil and facilitate the moisture entry into the soil.
- Tillage tools perform the smoothening and levelling process of the field.
- They break the clods and later pulverize them for efficient seed bed preparation.
- These tools cut the crop residue like straw etc and further mix them with soil for faster decomposition.

8.6.2 Checking equipments for maintenance

- Proper setup and maintenance of tillage equipment can eliminate extra tillage trips through the field and increase equipment efficiency, which preserves crop residue and limits soil erosion.
- Check for worn or broken tillage components, e.g., sweeps, chisel points, and disc blades. Damaged or worn cutting edges can increase tractor draft and result in an uneven soil surface. Wintertime is a good time to check for sales on replacement parts.
- Check the condition of the equipment's frame; look for sprung or broken welds and repair them. Also, see if the frame is bent because even a slight twist can result in a big difference in tillage uniformity from one end of the implement to the other.
- Look at the hydraulic system of the tractor. Is the hydraulic oil in good condition? Are the hoses all in good shape? Make sure that the weighting, ballasting, and condition of tractor are set up for an efficient drawbar pull.

8.6.3 Maintenance of Tillage Equipments

- ☆ Keep the secondary implements clean with regular checks.
- ☆ All nut bolts should be fully tight.
- ☆ The regular grease should be applied to the bearing after every set hours of operation.
- ☆ Tillers are one of the very basic secondary tillage implements used by the farmers and need apt care and preservation. To manage this implement better, you need to keep few things in mind:
- ☆ The teeth of tines should be kept at equal depth. To check this posture, place the tiller on plane ground and all the teeth should touch the ground. If the teeth are touching the ground, it means they are at equal depth.
- ☆ The teeth should have sharp edges so that good penetration can be maintained inside the soil. After the grinding process, the depth should again be checked.

8.6.4 Knowledge related to equipment use and maintenance

Moldboard Plows

- ☆ Provide adequate front-end weight for tractor stability in transport and operation, particularly with integral and semi-integral plows. Never pull from any point higher on the tractor than the recommended hitch point.
- ☆ Use extreme caution and reduce speed when transporting the plow and the tractor over rough ground.
- ☆ Avoid sharp turns at high speeds, especially on slopes.
- ☆ On tight turns, avoid swinging rear of plow into fences or other obstacles.
- ☆ Turning stops on semi-integral plows limit turning radius. Shorter turns may severely damage plow frame and tractor hitch.
- ☆ Never carry passengers on the tractor or permit others to ride on the plow -- particularly plows with automatic reset.
- ☆ Always lower the plow when not in use or left unattended.
- ☆ Lower the plow and securely pin the parking stand before detaching the plow from the tractor.
- ☆ Always use proper lighting, reflectors, SMV emblem, and other safety devices for road travel as required by state and local laws.
- ☆ When hitching drawn plows, always use a hitch pin with adequate strength for the tractor-plow combination.

Disk Plows

- ☆ Integral plows are transported completely raised, and all weight is carried by the tractor 3-point hitch. Adequate tractor front-end weights are required to offset the plow weight.
- ☆ When the transporting on a road or highway, always display SMV emblem and use lights and reflectors as required by state and local regulations.
- ☆ Semi-integral plows are quite long and caution must be used when turning to prevent swinging the plow into fences or irrigation ditches.
- ☆ Reduce speed when transporting over rough ground, and avoid quick, sharp turns at high speeds.
- ☆ When transporting semi-integral or drawn plows, always install cylinder locks to prevent accidental lowering of the plow. Relieve the load on hydraulic cylinders before starting to transport.
- ☆ Lower the plow to the ground or install hydraulic cylinder locks when the plow is not in use.
- ☆ Watch for other people when raising, lowering, or indexing the plow.
- ☆ Never permit anyone to ride on the plow, and allow only the driver on the tractor. Do not permit children to play on or near the plow either when parked or in operation.
- ☆ Lower the parking stand and securely pin it in place before detaching integral or semi-integral plows from the tractor.

Disk Tillers

- ☆ Keep SMV emblem clean and prominently displayed. Do the same with reflectors and warning lights as required by state and local regulations.
- ☆ Never allow anyone but the operator to ride on the tractor.
- ☆ Never ride or permit others to ride on the tiller.
- ☆ Lower the tiller to the ground when not in use.
- ☆ Secure the machine in the raised position by installing safety locks or hold-up pins when servicing or cleaning it.
- ☆ Disk blades are extremely sharp; be very careful when working or making adjustments in the disk area.
- ☆ Never walk close beside the rear wheel when the tiller is in operation. A sudden imbalance of forces could cause this wheel suddenly to jump to the left.

- ☆ Never grease, oil, or adjust the tiller while it is in operation.
- ☆ Escaping hydraulic oil under pressure can cause serious personal injury and infection. Therefore, be sure all connections are tight and that oil lines are undamaged. Always relieve hydraulic pressure in lines before disconnecting hoses. See a doctor immediately if escaping hydraulic oil has penetrated the skin.

Chisel Plow

- ☆ Reduce speed when transporting chisel plows over rough or uneven terrain.
- ☆ Use lock-up straps or transport locks when transporting a chisel plow.
- ☆ Be sure wings are locked in the folded position before traveling.
- ☆ Use proper lights, reflectors, and a clean SMV emblem when transporting equipment on road or highway.
- ☆ The transport width of most folding-section plows exceeds maximum width of normal vehicles. Therefore, use extreme caution when meeting other traffic, to avoid collisions and the possibility of transport wheels dropping into holes, drains, or ditches along the road edge.
- ☆ Allow only the operator to ride on the tractor.
- ☆ Never allow passengers to ride on the chisel plow.
- ☆ Never allow anyone to stand or work near the chisel plow when it is in operation, particularly when raising or lowering outriggers.
- ☆ Do not permit children to play on or near the chisel plow during operation or storage.
- ☆ Provide adequate tractor front-end ballast for stability in transport and operation, especially with integral models.
- ☆ Be particularly careful of escaping hydraulic fluid which can penetrate the skin and cause serious infection or reaction if not given immediate medical treatment.
- ☆ Do not unhitch from the tractor or store a chisel plow when outriggers are in the raised position.
- ☆ Make sure raised outriggers will safely pass under power and telephone lines.

Disk Harrows and Offset Disks

- ☆ Always lock safety lock during transport, if the disk is to be left raised for an extended period, or while working on the machine.

- ☆ Never depend on tractor hydraulic pressure to carry harrow weight in transport--use safety lock, and relieve pressure in cylinders.
- ☆ Lower integral harrows to ground each time tractor engine is shut off, and any time harrow is being serviced or repaired. If it must be raised for repairs, securely block the frame to prevent accidental lowering.
- ☆ Always use lights, reflectors, and SMV emblem when transporting, day or night.
- ☆ Lock the tractor drawbar in fixed position when transporting wheeled disks.
- ☆ Never transport a disk harrow on its own wheels at more than normal tractor speed, and considerably less than that on rough or uneven ground.
- ☆ Never clean, adjust, or lubricate the harrow while it is in motion.
- ☆ Wear protective gloves when working with or near disk blades.
- ☆ Hydraulic fluid escaping under pressure can penetrate the skin and cause serious infection or reactions. Never use hands to locate the source of a small leak which may be nearly invisible. Obtain immediate medical attention if injured by escaping hydraulic fluid.
- ☆ Park or block the harrow so it cannot roll when unhitched.
- ☆ Make sure wings are securely locked in transport position before moving the harrow.
- ☆ Large disk harrows exceed normal vehicle width, so be particularly careful to avoid collisions when meeting other vehicles on the road. Avoid dropping wheels of tractor or harrow into holes, drains, or ditches along the road.
- ☆ Provide adequate tractor ballast for front-end stability and to prevent excessive slippage.
- ☆ Never allow anyone to ride on tractor drawbar or harrow in operation or transport.
- ☆ Never allow anyone but the operator to ride on the tractor.
- ☆ Lower the machine or install safety lock when storing a disk harrow.
- ☆ Never permit children to play on or near a disk harrow while it is in operation, transport, or storage.
- ☆ Stand clear of harrow wings during folding or unfolding.
- ☆ Remove spring-loaded scrapers in proper order to avoid personal injury. Use care in relieving any spring under tension or compression.
- ☆ Do not make sharp turns with blades down.

Field Cultivators

- ☆ Never exceed recommended transport speed for the cultivator used; if speed is not stated, do not exceed maximum tractor speed.
- ☆ Reduce speed for turning and travel over rough or uneven ground.
- ☆ Use transport locks and relieve pressure in cylinders when transporting field cultivators. Do not depend on hydraulic pressure to carry the weight. Always lock wings in transport position and relieve pressure in cylinders.
- ☆ Never walk or work under wings when they are in the folded position.
- ☆ Follow state and local regulations regarding lights, reflectors, SMV emblem, and maximum width when transporting on roads or highways.
- ☆ Transport width of most field cultivators exceeds normal vehicle width. Therefore, use extreme caution when meeting other vehicles and avoid the possibility of dropping tractor or- implement wheels into holes, drains, or ditches along the road edge.
- ☆ Never permit anyone to ride on the tractor drawbar or cultivator in transport or operation, or to stand near the machine while it is operating--particularly when raising or lowering wings.

Toothed Harrows

- ☆ Provide sufficient front-end weight for safe, stable operation and transport.
- ☆ Use widest practical wheel tread to improve tractor stability, especially when working on steep slopes.
- ☆ Reduce implement to narrowest possible width for transport on roads or highways.
- ☆ Lock wings or folded sections securely in place before transporting equipment.
- ☆ Use lights, reflectors, and SMV emblem as required by law when transporting equipment--day or night.
- ☆ Schedule moves for least hazardous periods; avoid transporting equipment on busy roads, during peak traffic periods, or after dark.
- ☆ Never transport wheeled harrows, harrow carts, or field conditioners at more than tractor transport speeds; transport considerably slower on rough or uneven terrain.
- ☆ Never make sharp turns at high speeds.
- ☆ Never allow anyone to ride on the tractor but the operator.

- ☆ Never allow anyone to ride on the tractor drawbar or implement in operation or transport.
- ☆ Always stop implement and tractor engine to adjust, repair, or lubricate.
- ☆ Lower implements to the ground before stopping tractor engine, before servicing or repairing equipment, or at any time the machines are left unattended.
- ☆ Never make extremely short turns with drag-type eveners which could foul tractor tires.
- ☆ Never park implements where they could be hidden by tall or growing crops, grass, or weeds.
- ☆ Keep PTO shaft properly shielded. Never get off tractor without disengaging PTO and stopping engine.

Packers

- ☆ Always use reflectors, lights, and SMV emblem as required when transporting equipment--day or night.
- ☆ Provide adequate front-end weight for tractor stability in operation and transport of integral harrows. Use maximum allowable front ballast if operating integral harrows in lower gears.
- ☆ Never exceed normal tractor speed when transporting drawn roller harrows, and drive considerably slower than that on rough or uneven ground.
- ☆ Do not transport roller packers over hard-surfaced roads--use carriers.
- ☆ Never permit anyone but the driver to ride on the tractor.
- ☆ Never permit anyone to ride on the tractor drawbar or implement during operation or transport.
- ☆ Install transport lock pin before storing, transporting, or parking drawn implement; do not depend on hydraulic pressure to support the weight. Lower machines to the ground whenever the tractor engine is shut off.
- ☆ Pin tractor drawbar in center before transporting.
- ☆ Never lubricate, adjust, or repair the implement while it is in motion or the tractor engine is running.
- ☆ Always raise spring teeth before lowering roller harrow to the ground for parking or storage.
- ☆ Park or block the implement to prevent rolling when it is disconnected from the tractor.

- ☆ Never try to lift or support the roller harrow on the spring teeth for service or repairs.

Undercutters

- ☆ Use the SMV emblem, lights, and reflectors as required by law for transporting equipment on roads or highways.
- ☆ Install the transport safety lock and relieve hydraulic pressure in the cylinders when transporting.
- ☆ Be certain wings are securely locked in the folded position before transporting.
- ☆ Limit transport speed as recommended--15 miles an hour (24 Km/h) for some machines, even less on rough or uneven terrain.
- ☆ Never ride or allow others to ride on the machine during operation or transport. Allow only the driver on the tractor.
- ☆ Do not unhitch from the tractor or store the machine with wings in the folded transport position.
- ☆ Sweeps are sharp--watch out for them when wings are folded for transport or when servicing, adjusting, or repairing the plow. Always position wrenches to pull away from sharp edges or corners.
- ☆ Never stand with feet under blades while making adjustments or during maintenance. Be extremely careful while working within the implement frame.
- ☆ Do not stand or walk on the plow frame, or under wings when they are folded for transport.
- ☆ Stand with both feet on the same side of the tongue when hitching or unhitching.
- ☆ Do not allow children to play on or near the plow.

Rod Weeders

- ☆ Allow only the operator to ride on the tractor during operation and transport.
- ☆ Never permit anyone to ride on the rod weeder.
- ☆ Never exceed recommended transport speed or, if not stated, maximum tractor speed. Reduce speed on rough or uneven terrain or when turning.
- ☆ Use lights, reflectors, and SMV emblem as required by state and local regulations when transporting equipment, day or night.

- ☆ Avoid busy highways and peak traffic periods if possible. Move equipment only in daytime.
- ☆ Never attempt to repair, adjust, or lubricate the rod weeder while it is in motion.
- ☆ Shut off the engine and relieve the hydraulic pressure in hoses before disconnecting them.
- ☆ Never exceed recommended transport speed, or tractor road speed if maximum is not stated. Reduce speed when turning or crossing rough areas and slopes.
- ☆ Always lower parking stands on integral equipment before detaching from tractor.
- ☆ Always lower equipment or install transport lock before servicing, lubricating or repairing equipment, and when the machine will be left unattended.

Chapter 9

KNOWING ABOUT RULES, STANDARDS AND REGULATIONS

9.1 KNOWLEDGE ABOUT APPLICABLE LAWS, RULES, REGULATIONS, AND STANDARDS

9.1.1 Introduction

As with many laws that have been in effect for sometime, important points about the law are sometimes confusing, unclear or forgotten. Several points about regulations and application to agriculture need to be clearly understood by machine technician in agriculture. First, a technician employee relationship has to exist to apply to the business or operation. This means that if a farm operation uses only his or her own labor, or uses only family labor, there will no jurisdiction in that operation. The following are the main regulations which an agriculture mechanic should know related to farming:

9.1.2 Tractor Rollover Protection

- ☆ Also known as ROPS standard, ROPS have been required on all tractors operated by employees. As per the regulations, employers are required to provide safety operating instructions to employees when they start an assigned job and annually thereafter.
- ☆ Employers are to require seatbelt usage by employees when they are operating a tractor equipped with a ROPS.

- ☆ The only exemption to this standard is a low-profile tractor used in orchards, greenhouses, and other buildings.
- ☆ Training for employees should be comprehensive and cover topics about the tractor, safe operating procedures, and rules for tractor operation.

9.1.3 Accident Prevention Signs and Tags

- ☆ Such regulation defines the proper usage of the slow moving vehicle (SMV) emblem.
- ☆ The SMV should be constructed of the proper material and displayed appropriately at the rear of the tractor and/or tractor implement combination.
- ☆ The purpose of the SMV is to alert vehicle operators approaching from the rear that the farm vehicle is unable to travel faster than 25 mph.
- ☆ In regards to tags, an employer must provide lock out (LO) or tag out (TO) procedures to prevent accidental application of electrical power to farmstead equipment.
- ☆ Employee education is a required part of this standard to provide instruction in safe operation and servicing of all equipment that an employee will operate and general safety operational procedures.

9.1.4 Machinery Guarding

- ☆ Such standard requires that all farm field and farmstead equipment, regardless of date of manufacture, be equipped with power take-off (PTO) guarding.
- ☆ In addition to PTO guarding, equipment should be equipped with nip point guarding which guards against pinch points on gears, belts, and pulleys.

9.1.5 Worker's Compensation Laws

- ☆ Worker's Compensation laws provide financial help to workers injured on the job.
- ☆ An employer that employs a person in agriculture must provide worker's compensation coverage for that person is covered by the law, if the worker earns more than $1,200 in a calendar year, or if the employee works more than 30 days.
- ☆ Both the employer and employee pay into the WC fund of their home state based on their hours of employment.
- ☆ If an employee is injured, they must file a claim with their employers which will then be reviewed by approved providers.

- ☆ Compensation is dependent on the extent of the injury and medical check-ups may be required to determine the length of payments and a return-to-work date.

9.1.6 Hazardous Occupations Order

- ☆ The department of Labor has declared many agricultural tasks to be hazardous for youth under the age of 16.
- ☆ With certain exemptions, employment of youth under the age of 16 for these tasks is illegal.
- ☆ The law does not apply to youth under 16 who are employed, either with or without compensation, by their parents.
- ☆ As part of Labor Standards Act, hazardous occupations order in agriculture established a procedure where youth 15 and 17 years of age could be exempted from certain portions of the regulation.
- ☆ The exemption states that with successful completion of 10-hour training program, 15 and 16 year-old youths can be employed to "operate a tractor of over 20 PTO horsepower, or connect or disconnect an implement or any of its parts to or from such a tractor."
- ☆ With successfully completing a 20-hour training program, these youths can be employed to "operate or assist to operate (including starting, stopping, adjusting, feeding, or any other activity involving physical contact associated with the operation) any of the following machine":
    - Cornpicker
    - cotton picker
    - grain combine
    - hay mower
    - forage harvester
    - hay baler
    - potato digger
    - mobile pea viner
    - Feed grinder
    - crop dryer
    - forage blower
    - Power post hole digger
    - power post driver
    - or nonwalking rotary tiller

- ☆ With 10-hour training program, youths are allowed only to operate a tractor that essentially has no powered equipment hooked behind it.
- ☆ To do field work of any kinds, youths need to complete the 20-hour training program.
- ☆ In the definitions section of this law, the term "agriculture" is defined to include "farming in all its branches and among other things includes... preparation for market, delivery to market, delivery to storage or to market or to carriers for transportation to market."
- ☆ There are several tasks that are allowable for 15- and 16 year-olds to complete under the Fair Labor Standards including the following:
  - Loading and unloading trucks
  - Operating small tractors (under 20 horsepower)
  - Picking vegetables and berries
  - Placing vegetables and fruits on conveyors or into boxes
  - Clearing brush and harvesting trees up to 6 inches in butt diameter
  - Working with animals on the farm or at fairs and shows (except for specified breeding stock in confined areas, such as cows with newborn calves in closed box stalls, bulls, or sows with newborn piglets)
  - Raising and caring for poultry
  - Milking cows
  - Cleaning barns, equipment, and storage buildings
  - Mowing lawns
  - Riding, driving, or exercising horses
  - Picking cotton
  - Handling irrigation pipes
  - Riding on transplanters
- ☆ An employer may be strictly penalized if they subject a youth worker to hazardous occupations.
- ☆ Remember the youth is not penalized but the employer is the person held responsible with a first offense fine up to $10,000 and a second offense could include a fine and possible imprisonment.
- ☆ The HOOA does not permit youth younger than age 15 to complete the exemption training therefore they cannot be hired by an agricultural employer to operate tractors or machinery.

9.1.7 Worker Protection Standard

- ☆ Although not a farm machinery safety regulation the Worker Protection Standard (WPS) regulations of the Environmental Protection Agency (EPA) require employers to take steps to reduce the risk of pesticide-related illness and injury to those persons who use or are exposed to pesticides on farms, forests, nurseries, and greenhouses.
- ☆ No size of farm operation is exempt.
- ☆ The regulation requires the following:
  - Employer provides information to their workers about pesticide safety, emergency procedures, and recent pesticide applications.
  - Employer must properly train their workers about pesticide safety when they begin work and at least every five years.
  - Each handlers and early-entry workers must be provided with appropriate personal protective equipment.
  - In the event of an emergency, the employer must provide transportation for any employee who has been injured in a pesticide-related incident to an appropriate medical facility for treatment.
  - Employers must provide notifications (e.g., oral, written, posted) to alert employees about restricted-entry intervals.
  - Commercial handlers must provide the employer with complete details about the pesticide, warnings, and safety requirements prior to use.
  - Employers must provide an accessible and complete decontamination area within ¼ mile of all workers and handlers.

9.1.8 Vehicle Code

- ☆ The Vehicle Code includes several provisions that apply to the movement of agricultural equipment upon public roadways Of particular concern to agricultural producers are the definitions used in the code concerning implement of husbandry and highway.
- ☆ Implement of husbandry is a vehicle designed or adapted and determined by the Department of Transportation to be used exclusively for agricultural operations and infrequently operated or moved upon highways.
- ☆ Highway is the entire width between the boundary lines of every way publicly maintained when any part is open to the use of the public for purposes of vehicle travel.

9.2 WORK SITE PROCEDURES AND STATUTORY REQUIREMENTS

Policies, procedures and systems provide an invisible 'structure' to the business that helps everyone understand how and why things should be done. They can also serve to demonstrate safe and ethical business management through formal documentation. If procedures are well thought out and planned, they can help to make sure that everyone on the farm knows the safest and most productive way to complete their job.

Where appropriate, employees should be involved in the development of farm procedures relating to their everyday Work tasks. If employees are the ones completing these tasks on a daily basis they will likely have a better knowledge of what works and what doesn't. Including employees in the decision making also gives them a sense of 'owning' it and often makes them more willing to use it.

Work policies outline acceptable behaviors, set out consequences for not following policies, and often reinforce and clarify standard operating procedures. Agriculture Machinery Technician should develop workplace policies through following steps:

9.2.1. Consult employees

- ☆ Include the people working on the farm in the development and implementation of the workplace policies.
- ☆ Employees often have good ideas and including them in the development process will help to promote their understanding and ownership of the policies, awareness of the policy content and, importantly, increase the chances of compliance.

9.2.2. Draft policy document

A policy should:

- ☆ set out the aim clearly;
- ☆ explain why the policy was developed;
- ☆ define all the key terms in plain English so that the meaning is clear;
- ☆ list who the policy applies to, for example is it just people working on the farm or does it include contractors and visitors;
- ☆ describe what are acceptable or unacceptable behaviours (you may like to include examples help illustrate what is expected);
- ☆ have guidelines and procedures to be followed to achieve the policy aim;
- ☆ set out the consequences of not complying with the policy and, if necessary, what to do in the case when it isn't possible to comply;
- ☆ provide a date of when the policy was developed or updated and;

9.2.3. Discuss with staff and people

- ☆ For farm workplace policies to be effective, tell everyone who works on the farm about them and give them a copy.
- ☆ If you have any employees whose first language is not English and whose understanding of written English is limited, you should arrange to have the policies translated into the appropriate language. And don't forget visitors including employees' children's friends. It is important that everyone understands the policies.
- ☆ Similarly, if you have employees who have limited literacy skills you should ensure that the policies are read to them and understood by them.
- ☆ Contractors, labour hire firms and work experience students also need to be aware of your policies and it may be necessary to consult with them to ensure consistency and compliance with their policies and procedures.
- ☆ Make copies of the policies easily accessible on the farm by keeping copies in folders in a central location and in staff areas and displaying them on noticeboards.
- ☆ Make sure that policies are provided to new staff, contractors and, where relevant, visitors as a part of their induction.

9.2.4. Explain the policies

- ☆ Explaining the policies to everyone on the farm can be done at information or training sessions, staff meetings, or when a new staff member or contractor is completing their induction.
- ☆ Explain how to comply with the policies and the implications of not doing so.
- ☆ Make sure that everyone understands they must comply with the policy.
- ☆ Regularly discuss policies with the people working on the farm. This could be done at staff briefings or meetings whenever relevant.
- ☆ The best way for managers and owners to show others working on the farm how to behave is by modelling behaviours in line with the farm policies.

9.2.5. Implement the policies

- ☆ Be consistent when applying the workplace policies and dealing with someone who doesn't comply.
- ☆ Deal with any breaches of policies promptly and always use the procedures that are set out in the policy.
- ☆ The consequences of breaching a policy could be a warning, disciplinary action or dismissal.
- ☆ The consequence should be suited to the severity of the breach.

9.2.6. Evaluate and review the policies

- Regularly review the policies in conjunction with employees and contractors to make sure that they match any changes that have occurred on the farm and reflect new regulations and state of knowledge so they remain relevant to the people who work there.
- Policies will also need to be updated in line with any changes to workplace laws or regulations.
- If a policy is changed significantly give a revised copy to all staff and update any other copies that are on the farm.
- Explain to everyone why it has been changed and what the changes are. Policies may need to be changed when there is new equipment being used or work procedures change.

9.3 LOCATING HAZARDS AND TAKING ACTIONS AT WORKPLACE

There are many types of harzards which occurs at workplace due to:

9.3.1 Children

Children who live on farms or who come to visit are often at greater risk than adults who work there. To make your farm safer for children, hazards must be spotted and risks minimised before children discover them. The safest farms for children are those where safety is a priority for everyone.

Identify places where children like to play, perhaps where they are not supposed to be, and the sort of things they might like doing.

Consider dams, streams and pools, silos, tractors, electricity, workshops and machinery sheds, chemical storage areas, farm bikes, guns and dangerous stock.

For each identified risk, assess the likelihood and possible severity of injury or harm.

Ask the children to help. Make high risk areas your top priority for safety improvements.

For minimise risks to children on workplace:

Fences

- For small children, have an effective fence around the house and yard.
- Fence off septic tanks, sheep dips, seepage pits, ponds dams, pools and creeks particularly if close to the house.
- Maintain fences round nearby paddocks and work yards to protect small children from animals, vehicles, machinery, road traffic.
- Have safe, fenced-off areas where children can play.

Workshop

- ☆ Ensure gates, doors and locking systems keep young children out of workshops and hazardous storage areas.
- ☆ Have safety rules for older children who may need to enter these areas on farm duties.
- ☆ Keep workshops free from child hazards relating to electricity, power tools, fire, poisoning, slips, trips, falls and other dangers.

Pesticides

- ☆ Keep farm pesticides locked away out of children's reach.
- ☆ Fence off pesticide mixing and wash-down bays to prevent access by children.
- ☆ Keep children out of orchards after spraying.

Silos, grain storage

- ☆ Keep grain storage bins, silos, augers and trucks, adequately guarded to prevent access by children.
- ☆ Never allow children to play on stored grain in silos.
- ☆ Ensure fixed ladders are guarded and kept above children's reach.
- ☆ Have rules keeping children out of grain loading and storage areas unless under close supervision.

Machinery and equipment

- ☆ Lock tractors, trucks and other farm machinery away after use, out of bounds to children.
- ☆ Electrical appliances and tools should be turned off, disengaged and kept inaccessible to young children.
- ☆ Keep firearms, ammunition and explosives locked and out of children's reach.

Protection from animals

- ☆ Have rules to safeguard children from dogs that might attack or bite.
- ☆ Ensure small children cannot wander into animal pens and stockyards with confined stock.

Ladders

- ☆ Store ladders away to prevent children climbing roofs, silos, trees and other height hazards.
- ☆ Ensure fixed ladders on silos, bins, tank stands, windmills etc. are adequately guarded against children attempting to climb them.

Emergency first aid

- ☆ Have an emergency plan for dealing with serious accidents.
- ☆ Keep a first aid kit suitable for children, and have someone trained in first aid.

9.3.2 Dairy Farms

Dairy farmers often work in isolation, facing risks from animal behavior, mechanical hazards, climatic conditions, and rushed work deadlines.

Look for hazard related to lighting, electricity, slips and trips, training and supervision of new and young workers, animal behavior, machinery guarding, heavy lifting and carrying.

Check each identified hazard for likelihood and severity of injury or harm. The greater the risk and severity, the more urgent it is to minimise or eliminate the risk. Consider appropriate changes and make sure new hazards are not created.

Minimise risks in dairy farming by:

- ☆ Have adequate lighting for early morning and evening milking.
- ☆ Concrete surfaces should be roughened to provide extra traction for both handlers and stock.
- ☆ Keep guarding in place on moving parts, e.g. belts and rotaries.
- ☆ Check guarding on compressors, pumps, electric motors and grain augers.
- ☆ Have an emergency stop lanyard - in addition to the forward-stop-reverse lanyard.
- ☆ Have a residual current device (RCD) installed on the electrical circuit board.
- ☆ Fit all-weather covers on power boards in wet areas.
- ☆ Ensure milk line supports and union joints meet recommended safety levels.
- ☆ Cover head-high projections like handles on milk filter casings with padding.
- ☆ Keep exhaust pipes clear of walkways.
- ☆ Maintain exhaust systems in good order to reduce noise and fumes.
- ☆ Fence off effluent disposal ponds to keep out children and stock.
- ☆ Clearly mark all water outlets not suitable for human consumption.
- ☆ Ensure hot water taps are inaccessible to children.

Strain injuries

Activities that can lead to back strain injuries include:

- ☆ long hours working on tractors;
- ☆ stock feeding;
- ☆ fencing;
- ☆ hay and silage preparation;
- ☆ irrigation.

Reduce risk of back strain injuries,

- ☆ use mechanical aids, such as hoists, trolleys, barrows and pulleys;
- ☆ use team lifting, planning each task in advance;
- ☆ keep loads small;
- ☆ keep walkways clear;
- ☆ modify work areas to minimise bending, lifting, pulling, pushing, lowering and carrying.
- ☆ do repetitive tasks at a comfortable height, with the least amount of bending, stretching or leaning.
- ☆ develop safe lifting techniques - using the legs and not the back.

Hot water

- ☆ Ensure hot water is safely guarded.
- ☆ Have safe procedures for working with or near hot water.
- ☆ Make sure hot water taps can be clearly identified.
- ☆ If appropriate, fix clear warning signs next to hot water hazards.

Remember

- ☆ Ensure adequate lighting for milking.
- ☆ Plan tasks and modify equipment to minimise hazardous manual handling.

9.3.3 Electricity

The best safeguard against electrocution on farms is the residual current device (RCD) or safety switch. A fixed RCD can be installed instead of a fusebox in your house, shed or workshop, or portable RCDs can be used with individual power tools.

Check to ensure electrical fittings, fixtures, plant and equipment, wiring, insulation, switches, power cords, plugs, earth wires, guarding, and welding equipment are in good condition and regularly maintained.

Look for shorting or sparking fittings. Avoid using electrical equipment in wet conditions. Wear safe footwear and clothing. For work on wires, plugs, switches, fuses and electrical plant, call the electrician.

Assess each identified hazard for likelihood and severity of possible injury or harm. If there is any risk of electric shock or electrocution, you should have a safe procedures to ensure the hazardous plant is put out of use and either isolated, or kept in a safe place until repaired or discarded.

Minimise risk of electric shock:

- Make sure all hand held power tools and appliances are connected through an RCD.
- Always employ an electrician for power alterations or repairs.
- Ensure wiring, equipment, leads and plugs are kept in good repair.
- Don't overload your wiring installation.
- Don't remove guards or covers from electrical switch gear.
- In areas exposed to wind and rain, always use weather-proof outlets and fittings.
- Avoid using outdoor electrical equipment in wet weather.
- All lights exposed to breakage by farm tools should be fitted with wire guards.
- Old rubber-insulated wiring is now unsafe, and should be replaced.

Earth wires

- The earth wire is an essential safety feature. Its purpose is to divert any current leakage to the ground and cause a fuse to blow or an RCD to trip out should a fault develop in the installation.
- The earth wire is usually a bare or green and yellow insulated copper wire, connected to a water pipe or stake driven into the ground. It should never be removed or disconnected.

Outdoor power lines

- Make sure tall items like balers and headers are kept well clear of overhead wires.
- Never ride on top of high loads.
- If your crop-dusting is done by aeroplane, tell the pilot beforehand about any power lines in the area.
- Plan farm roads to avoid passing under power lines, and have new power lines installed so they don't cross over roads.

- ☆ Always check the location of power lines before you start work.
- ☆ Always check plans and records of underground powerlines before any digging or earthworks.
- ☆ Never stack irrigation pipes or park machinery under power lines.
- ☆ Never up-end a pipe before looking up. Carry pipes horizontally.
- ☆ Remember, power line heights are deceptive. Know the operation and maximum height of your machine.
- ☆ Have an observer check your position when working close to overhead power lines.
- ☆ If in doubt, always contact the supply authority for advice and assistance.
- ☆ If you see a power line that has been damaged or has fallen down, keep clear and notify the supply authority.

Fuses and RCDs

- ☆ If a fuse blows out, turn off the switch and check the electrical equipment being used before you replace the fuse wire. If the fuse blows again, call an electrician.
- ☆ When replacing a fuse wire, make sure its rating is correct for the circuit.
- ☆ If an RCD trips out, check the electrical equipment for obvious faults. If it keeps tripping out, call an electrician.
- ☆ Remember, while an RCD may shut off a lethal dose of electricity, it does not prevent electric shock. You must still avoid live contact, particularly if you are working at height or operating hazardous machinery.

Power tools

- ☆ When buying a portable power tool, double insulated is safer.
- ☆ Never use a light socket to operate a power tool.
- ☆ Don't use tools if the casing is broken or damaged. Damaged cords and plugs should be re placed.
- ☆ Regularly check power tools are free from external damage or makeshift repairs. This includes leads and plugs.
- ☆ Don't make adjustments to a tool without first switching it off and removing the plug from the power point.
- ☆ All bench-mounted equipment, such as power saws or grinders, should be effectively earthed - except for those with double insulation.

- ☆ Don't use makeshift extension lights. Use a type with a guard around the globe and an insulated handle.

Welding equipment

- ☆ Switch off power before connecting welding leads to terminals.
- ☆ Check leads are correctly connected to terminals marked 'electrode' and 'work'.
- ☆ Ensure supply terminals and live parts are suitably enclosed and protected.
- ☆ Ensure welding terminals are shrouded to prevent inadvertent contact or short circuits.
- ☆ Check the frame of welding equipment is effectively earthed.
- ☆ Don't use leads if they have bare sections. Replace them.
- ☆ Never use bare hands on metal parts of electrode holders or electrodes while the welder is switched on. Never rest the electrode on your body.
- ☆ Be sure to keep waste material away from the welder.

9.3.4 Farm chemicals: Pesticide Spray

Most pesticide sprays are hazardous, and can cause injury or harm if not handled and applied correctly. Use them in accordance with labels supplied on containers, and with the more detailed material safety data sheets (MSDSs), available from your supplier.

Read labels and MSDSs carefully. Check spray equipment, safe handling procedures, protective equipment, operator training and awareness, and supervision of new and young workers. Be aware pesticides enter the body through:

- ☆ absorption through skin contact, specially the eyes;
- ☆ inhalation of fumes, vapors and dusts;
- ☆ accidental swallowing while eating, drinking or smoking.

Assess the effectiveness of protective equipment, decanting, spray and washdown procedures, operator training and safe practices. Assess also the likely risk of contamination or poisoning occurring, and the severity of harmful effects to the operator. Finally, assess the effectiveness of proposed new safety measures.

Here are some ways of improving farm spray safety:

- ☆ Use the least toxic pesticide available for effective control of insect, fungus or plant, as the case may be.
- ☆ Ensure only the recommended rate of pesticide is used.

- ☆ Wear protective clothing and equipment as described on the label and MSDS.
- ☆ Prepare only enough chemical for immediate use.
- ☆ Keep a record of each use and the results.
- ☆ Ensure equipment works properly and does not leak.
- ☆ Cover feed and water containers near areas where livestock are grazing;
- ☆ Don't eat, drink or smoke while pouring, mixing or spraying.
- ☆ Don't pour concentrated pesticide into tanks above shoulder height.
- ☆ Avoid working alone if you are using a highly toxic pesticide; or have some form of mobile communication.

Spraying

- ☆ Spray with minimal drift and preferably in low wind conditions.
- ☆ Never spray in high wind conditions.
- ☆ Use mechanical suction to transfer pesticides to spray tank.
- ☆ A vortex system can be used to mix pesticide concentrate with water before filling the spray tank.
- ☆ Prevent nozzles blocking by using correct filters and pesticide formulation. Ensure water and equipment are clean.
- ☆ Clear blocked nozzles by using a soft bristle brush or compressed air. Never suck or blow nozzles to clear them.

Clean up

- ☆ Thoroughly clean all spraying and protective equipment, where run-off will not create a hazard or contaminate the environment.
- ☆ Wash work clothing separately from domestic clothing, or use disposable clothing.
- ☆ Wash yourself well after a spray operation.
- ☆ After handling pesticides, wash hands with soap and water before eating, drinking, going to the toilet or smoking.

Take precautions

- ☆ Provide a first aid kit that includes a towel, clean clothing, a resuscitation mask for expired air resuscitation, disposable eye wash bottle and eye wash solution, soap, nailbrush, and clear instructions on what to do with this equipment.
- ☆ Keep fresh water close by for washing.

- Advise someone where you will be working and how long you intend to be gone. Otherwise ensure you have a two-way radio for emergency calls.
- Have a family member or other person at the farm attend first aid classes.
- Check the labels, MSDSs or other safe handling guides for your pesticides, to ensure you have the correct antidotes, emergency equipment and facilities required by labels and MSDSs.
- Stop work immediately and seek medical attention if there is any sign of muscular spasm, blurred vision, excessive saliva or difficulty breathing. Suspect pesticide poisoning with any of these symptoms.
- If regularly using pesticides, an annual medical examination is recommended before and after the spraying season, and more frequently with some chemicals.

For skin contact: Wash with soap and water, and rinse with clean water. Remove any contaminated clothing, and seek medical advice.

For eye contact: Hold eye open under running water for 15 minutes. Seek medical advice.

For swallowing: Ring the Poisons Information Center .

9.3.5 Farm Noise

Noise from farm tools and machinery can cause permanent hearing loss. Hearing loss may be temporary at first, but repeated exposure will lead to permanent damage. The damage can occur gradually over a number of years and remain unnoticed until it is too late. Some noises, such as gunshots, are so loud they can cause immediate permanent damage.

Some early warning signs of hearing loss include:

- ringing in the ears after work;
- difficulty understanding a normal conversation;
- turning up the volume on radio or television when others appear to hear adequately;
- failing to hear background noises, such as a ringing telephone or doorbell.

Typical farm noises that can damage hearing include:

- tractor (95-100dB(A))
- header (88-90dB(A))
- angle grinder (95-105dB(A))

- ☆ bench grinder (90-95dB(A))
- ☆ chainsaw (105-120dB(A))
- ☆ shotgun (over 140 dB(lin)).

If you have to shout above noise to be heard by someone a metre away, your hearing could be at risk. If noise cannot be reduced or removed at its source, and if there is no other way to separate people from damaging noise exposure, protective hearing equipment must be worn. Some farmers employ a noise consultant to take noise readings, assess hearing risks and recommend preventive measures.

You can reduce noise at its source by:

- ☆ purchasing quieter machinery and equipment;
- ☆ modifying equipment to reduce noise and keeping machinery well maintained;
- ☆ if practicable, running machinery at lower revs.

You can protect people from loud noise exposure by:

- ☆ limiting the time workers spend in a noisy environment.
- ☆ isolating work areas from noisy machinery using distance or insulation;
- ☆ scheduling noisy work when fewer workers are around;
- ☆ using job rotation to alternate noisy jobs with quiet ones.

Protective equipment

- ☆ Where noise exposure cannot be reduced, hearing protection should be worn, e.g. on open tractors, when shooting, or when using a chainsaw.
- ☆ Try on ear muffs before buying, to ensure comfort and a sound-proof fit.
- ☆ The higher the SLC 80 (sound level conversion) figure, the higher the protection.
- ☆ Use lower SLC 80 muffs for moderately noisy jobs - a high rating might mask out important danger warning sounds.
- ☆ Ear plugs may be more comfortable for some farmers, but must be inserted with clean hands. Re-usable plugs must be cleaned regularly. Cotton wool is not sufficient.
- ☆ Clean and maintain hearing protectors. Replace worn or damaged parts. Keep protectors near the area of noisy activity, e.g. in the tractor cab.
- ☆ Wear a combination of ear muffs and ear plugs when shooting.

9.3.6 Fire Fighting

Teamwork, planning and communication are vital to the safety of people fighting fires, burning off or doing other fire prevention work.

The main hazards are smoke inhalation and radiant heat. Associated hazards relate to training and safe fire fighting practices, communication between firefighters, other people involved, wind and weather conditions, terrain and vegetation, threatened buildings and their contents, availability of water and fire fighting machinery. People on farms should know how to contact the area fire control officer.

Safe fire fighting procedures involve constantly assessing risks and their potential to endanger life and property, and minimising them where possible.

The following safety rules help firefighters minimise risks.

- ☆ Never work alone.
- ☆ Anticipate fire changes due to wind, topography and fuel type.
- ☆ Watch for erratic fire behaviour.
- ☆ Beware of burning limbs and trees in previously burnt country - look up and live.
- ☆ Keep clear of all vehicles or machinery - the operator may not see you.
- ☆ Avoid steep slopes above a fire.
- ☆ Observe and keep in mind local topography:
- ☆ position of tracks, clearings, creeks and other relevant spots or landmarks
- ☆ avoid danger areas like steep slopes, dense vegetation and deep, narrow gullies.
- ☆ Select escape routes before entering the fire zone.
- ☆ Obtain the latest forecast with particular attention to wind changes.
- ☆ Remember a general forecast may not apply in your area due to fire effects, terrain or local factors
- ☆ Relate local weather to possible fire behaviour.

Personal effort

- ☆ Maintain self control under threat situations.
- ☆ Panic is infectious and drains energy.
- ☆ Avoid exhaustion from over-exertion or prolonged periods of effort.
- ☆ Avoid unnecessary shouting or whistling - it may confuse others.

Limits of endurance

- ☆ Lack of rest reduces physical strength, the ability to think clearly and speed of reactions - arrange relief for yourself and your team.
- ☆ Take a break at every opportunity.
- ☆ If feeling ill, drowsy, faint or nauseated, take action for heat stress.

Drinking

- ☆ To do without water for long periods is a mistake. Dehydration can make you sluggish, irritable, impatient, muddle-headed, tired and sleepy.
- ☆ When sweating freely, replace body fluids and salts. Drinking small quantities frequently is better than having one big gulp every hour.
- ☆ Replace salt by taking salt in food and drink, e.g. Vegemite or lightly salted water (one level teaspoon per litre).
- ☆ Aerated drinks blow up the stomach and make hard work uncomfortable. Beer does not help either.
- ☆ No alcoholic drinks. Wait until the fire is finished before 'one with the boys'.

Protective clothing

- ☆ Guard against falling objects - wear an approved safety helmet.
- ☆ Make sure your helmet is properly adjusted. In severe conditions, wear a chin strap.
- ☆ Wear safety glasses, goggles or a face shield to prevent injury from windblown dust, smoke irritation or during chainsaw operations.
- ☆ Working boots must be in good condition. Wear approved safety boots.
- ☆ Woollen clothing offers some protection against fire.

Heat and smoke

- ☆ Protect against radiant heat.
- ☆ Take refuge:
  - light a back burn and use burnt country as a refuge;
  - use gravel pits, or clearings in the forest and roads;
  - lie down on the ground - air is freshest and coolest at ground level.
- ☆ Use vehicles to shelter from radiation when the temperature becomes uncomfortable.
- ☆ Don't take refuge in elevated water tanks. Immersion in lukewarm water can kill.

- ☆ Limit breathing rate when smoke is dense - wait for small pockets of fresh air.
- ☆ Dense hot smoke could damage lungs, but hot dry air can be breathed for some time without lung damage.
- ☆ If it is necessary to move through the flames:
  - don't linger in front of the flames;
  - use clothing to the best advantage as a shield;
  - select an opening where flame height is lowest;
  - move through the flames onto burnt ground as quickly as possible;
  - beware always of the danger from falling limbs and trees in burnt country.
- ☆ As a last resort, if trapped, lie on the ground taking advantage of any protection available.

9.3.7 Handling Cattle

Injuries from cattle relate to a number of factors - inadequate yard design, lack of training of handlers, unsafe work practices, and the weight, sex, stress factor and temperament of animals.

- ☆ Check accident records to identify tasks most likely to cause injury.
- ☆ Consider situations that cause stress and injury to handlers and stock.
- ☆ Take into account sex, weight and temperament of stock.
- ☆ Consider effects of weather and herding on animal behaviour, and time allowed for settling down.
- ☆ Check potential hazards and safety advantages of stock facilities, including mechanical aids and work layout.
- ☆ Consider what training is required before a person can confidently and competently handle stock.

Risk

- ☆ Using accident records, check which tasks and work situations are most frequently linked with injuries.
- ☆ Discuss safety concerns of handlers in regard to various tasks.
- ☆ Check each identified hazard for likelihood and severity of injury.
- ☆ Assess proposed safeguards and safe procedures for other hazards.

Improving safety in cattle handling

- ☆ Always plan ahead. Prepare and communicate safe work practices. Get assistance if necessary.

- ☆ Wear appropriate clothing, including protective footwear and a hat for sun protection.
- ☆ Make use of facilities and aids - headrails, branding cradles, whips, drafting canes, dogs etc.
- ☆ Know the limitations of yourself and others - work within those limitations.
- ☆ Respect cattle - they have the strength and speed to cause injury.

The stock

- ☆ Hazards vary according to the age, sex, breed, weight, horn status, temperament and training of animals.
- ☆ Approach cattle quietly, and make sure they are aware of your presence.
- ☆ Bulls are more aggressive during mating season and extremely dangerous when fighting. Separate into different yards where appropriate.
- ☆ Cows and heifers are most likely to charge when they have a young calf at foot.
- ☆ Heifers can also be dangerous at weaning time.
- ☆ Isolated cattle often become stressed and are more likely to charge when approached.
- ☆ Cattle with sharp horns are dangerous - dehorning is recommended where practicable. Dehorned and polled cattle can still cause injury.

Hay Baling

Large hay bales, some weighing up to 800 kg, have killed and seriously injured many farm workers. Bales, both round and rectangular, can fall on tractor and forklift operators , topple off stacks and vehicles on workers or bystanders, and collapse when stacks fail.

Look for hazards relating to:

Handling:

- ☆ Children playing near hay balers, carriers and stackers.
- ☆ Training of operators handling tractors, front-end loaders or forklifts.
- ☆ Two-poster tractor ROPS - offer no operator protection from bales falling back off forks or bale-loading frames. Tractors with cabs, FOPS (fall on protective structure) or four-poster ROPS are safer.
- ☆ Makeshift or poorly fitting bale-loading attachments on tractors and forklifts.

- ☆ Carrying bales too high off the ground.
- ☆ Insufficient counterbalance on tractor or forklift vehicle.
- ☆ Hydraulic control valve should be specific to the front-end loader attachment.

operation:

- ☆ Baler properly connected to the tractor.
- ☆ Adequate safety guards fitted.
- ☆ Nobody allowed to ride on the baler.
- ☆ Prevent others getting too close to the baler.
- ☆ Build-up of loose, combustible material in the baler.
- ☆ Fire extinguisher fitted to the machine.
- ☆ Disengage PTO and apply fly wheel brake prior to making baler adjustments.
- ☆ Stop engine and apply fly wheel brake before repairs or "stringing up" the baler.
- ☆ Extra care and attention when reversing or turning the machine,
- ☆ working at night,
- ☆ loading onto a truck, and unloading.

Check each hazard that has been spotted to assess:

- ☆ Likelihood - how likely is this hazard to injure someone? and
- ☆ Severity - how severe would that injury be?

Handling:

- ☆ Keep children away from hay baling and stacking operations.
- ☆ Make sure operators and handlers are properly trained and physically capable of tasks.
- ☆ Consider fall-arrest protection for people working at heights.
- ☆ Use tractors with four-poster ROPS, FOPS or cabs for protection against falling bales.
- ☆ Replace risky attachments with manufacturer approved attachments.
- ☆ Ensure loader or forklift operators transport bales close to the ground.
- ☆ Avoid sharp turns and unsafe speeds.
- ☆ Make sure vehicle controls are fitted specifically for the attachment in use.

Stacking:

- ☆ Stacks should be on firm, level ground, away from fire hazards, sources of ignition, overhead powerlines, dwellings, boundary fences and footpaths.
- ☆ Make sure stack and load heights do not exceed the lifting capabilities of the farm handling equipment.
- ☆ Big bales should be stacked to a maximum of four bales high.
- ☆ High density bales can be stacked up to six layers high.
- ☆ Stack big rectangular, square or high density bales by overlapping, to form a stable stack.
- ☆ Do not use bale lifting equipment to raise people on or off stacks.

Transporting:

- ☆ Do not allow people to ride on stacked trailers.
- ☆ Be aware of overhead obstructions, like trees, bridges and powerlines.
- ☆ Avoid rough ground that could cause bales to dislodge.
- ☆ Ensure loads are adequately secured. Do not overload vehicles beyond legal limits.

9.3. 8 Heat Stress

The effects of heat stress range from simple discomfort to life threatening heat stroke. Heat stress causes increased sweating which leads to loss of body fluid and then reduced heat tolerance. This results in reduced capacity for work, inefficiency, and increased risk of hazardous incidents.

Heat stroke, a rarer condition, is when sweating stops and body heat rises. This is a life threatening condition, and requires immediate medical attention.

Heat stress hazards can occur through:

- ☆ high temperatures,
- ☆ high humidity,
- ☆ lack of air movement,
- ☆ unsuitable clothing,
- ☆ a person's lack of acclimatisation,
- ☆ hot protective clothing or equipment,
- ☆ physical activity,
- ☆ radiant temperature of surroundings.

Using weather forecasts, availability of shade, knowledge of the job ahead, and an awareness of individual workers' heat tolerance, assess whether the day's tasks could cause heat stress or heat stroke. Consider ways of minimising or eliminating the risks.

Working in a hot environment is more likely to adversely affect people who are:

- ☆ overweight,
- ☆ medically unfit,
- ☆ unacclimatised to heat,
- ☆ unhealthy, particularly if suffering from heart disease,
- ☆ dehydrated, whether from alcoholic hangover, failure to replace salt and water lost in sweat, or from medically prescribed diuretic drugs.

Indoors

- ☆ Open windows and doors to allow natural cross ventilation, or install air conditioning if practicable.
- ☆ Provide fans or ventilators to lower temperature and increase air movement.
- ☆ Insulate roof, walls or heat making equipment.
- ☆ Duct hot steam and gases outside to help reduce humidity and lower temperature.
- ☆ Install extraction ventilation around heat producing equipment.

Outdoors

- ☆ Wear cool cotton clothing to allow air circulation and evaporation of sweat.
- ☆ Use a broad spectrum sunburn cream with an SPF of 15+.
- ☆ Wear broad brimmed hats that shade head, neck, face and ears.
- ☆ Wear close fitting sunglasses with side shields, labelled to meet Australian Standard AS 1067.
- ☆ Use a wetted scarf.
- ☆ Provide shaded rest areas.
- ☆ Provide an ample supply of cooled, non-alcoholic drinks and ensure they are easily accessible.
- ☆ Drink small amounts of water at frequent intervals to avoid dehydration.
- ☆ Re-schedule heavier work for cooler times of the day or for cooler days.
- ☆ Where possible, rotate work so workers spend less time each on heavy tasks.

9.3.9 Machinery Guarding

Manufacturers of machinery and equipment are today legally required to make sure dangerous parts are safely guarded so that operators and others are protected from injury.

But old farm machinery is sometimes poorly guarded. Extra moving parts like wheels and pulleys may have been added for various other uses. Original guarding may have been removed for maintenance and not put back.

There may be times when an operator may need to reach over, under, around or into a machine while it is running. If so, any moving parts or other hazards must be appropriately guarded from human contact.

A guard may be any shield, cover, casing, or physical or electronic barrier, intended to prevent contact between a hazardous machine part and any part of a person or a person's clothing.

Some of the hazards associated with machinery likely to cause injury include:

- ☆ rotating PTO and other shafts (e.g. joints, couplings, shaft ends and crank shafts);
- ☆ gearing (incl friction roller mechanism), cables, sprockets, chains, clutches, cams or fan blades;
- ☆ the run-off point of any belt, chain or cable. All belts are hazardous, especially if joints are not kept smooth.
- ☆ keyways, keys, grease nipples, set-screws, bolts or any other projections on rotating parts;
- ☆ any pulley or flywheel that incorporates any openings, spokes, protrusions, etc, that render it anything except totally smooth;
- ☆ any crushing or shearing points, e.g. augers and slide blocks, roller feeds, conveyor belts;
- ☆ ground wheels and track gear that incorporate protrusions, spokes, etc, that are adjacent to an operator's position (standing platform, seat, footrest) or passenger's seat;
- ☆ rotating knives, blades, tines or similar parts of power driven machines that operate in or near the ground or engage crops;
- ☆ any machine component that cuts, grinds, pulps, crushes, breaks or pulverises farm produce;
- ☆ hot parts of any machine where the surface temperature exceeds 120C in normal operation.

Once a hazard has been identified, assess the likelihood of the hazard resulting in injury to the operator or any other person, and the likely severity of any injury or harm.

Ensure machinery guards:

- ☆ are designed to protect the user but allow ready access for operation and maintenance;
- ☆ are always in place on dangerous parts of machinery unless they are, by any reasonable definition, located out of reach of users, operators or bystanders;
- ☆ are conveniently placed so that users, operators and service and maintenance people are less likely to remove them permanently;
- ☆ are strong and durable enough for the machine part they cover;
- ☆ protect users, operators and bystanders against burns caused by hot parts;
- ☆ are ventilated where applicable to avoid the machine over-heating;
- ☆ are not removed until the machine is stopped and isolated with a tagged lock-out switch, and all sources neutralised, e.g. pressure in the hydraulic, or LPG gas line.

Children on or visiting the farm are often at risk of being injured by machinery. Minimise the risks, teach your children about safety on the farm, and get them to tell their friends.

- ☆ Agricultural machinery is not a playground. Make sure guards are on machines, especially when children are around;
- ☆ Be aware that children's fingers can sometimes reach into guarding designed for adult hands.

Stick to a safe procedure for machinery guarding.

- ☆ For maintenance jobs, have a checklist procedure ensuring guarding is safely replaced.
- ☆ Use approved lock-out and tag devices to prevent machinery being accidentally started during maintenance.
- ☆ Redesign work processes to minimise risk from moving parts.
- ☆ Get rid of machinery and eliminate work processes that can't be made safe.
- ☆ Replace unguarded machinery with safer machinery.
- ☆ Have guards designed and fitted for improvised machinery.

Manual Handling

Manual handling or strain injuries can keep farm workers away from work for weeks at a time. They can happen from lifting, pushing, pulling, carrying, lowering, holding or restraining.

Injuries occur through:

- ☆ increased wear and tear or damage, e.g. from intense or strenuous manual activity;
- ☆ gradual wear and tear, e.g. from frequent or prolonged periods of activity (continuous handling of hay bales); heavy or awkward lifts (lifting heavy machinery onto a ute);
- ☆ sudden damage, e.g. from unexpected movement (carrying a heavy object over uneven ground, stumbling, tripping or falling).

Conduct safety audits of all farm jobs involving manual handling. Take note of heavy, stressful, awkward or repetitive activities. Check injury records to see which activities have caused most strain injuries. Look for difficult handling jobs that could be made easier.

Assess the likelihood of each identified hazard resulting in injury or harm. Use injury records to assess the potential risk of various tasks. If you consider there is a significant risk of serious injury, look for the best way to minimise the risk.

Here are some suggestions to help you make the changes:

- ☆ Plan ahead. Consider the safest possible ways of lifting, carrying, holding, lowering, pushing, pulling.
- ☆ Eliminate unnecessary tasks.
- ☆ Avoid double handling.
- ☆ Use mechanical aids.
- ☆ Carry out a safety check first.

Lighten the load

- ☆ Where possible, choose light-weight materials.
- ☆ Divide heavy loads into smaller loads.
- ☆ Purchase in smaller bags.
- ☆ Half fill containers.
- ☆ Get help to share the load.

Reduce bending, twisting, reaching

- ☆ Point your feet in the direction of the load you are carrying.
- ☆ Keep tools and equipment within easy reach.

- ☆ Build benches to waist height.
- ☆ Keep frequently used items at waist height.

Follow a safe procedure

- ☆ Plan the handling.
- ☆ Clear the way.
- ☆ Wear appropriate protective clothing.

Correct body techniques

- ☆ When lifting a load from ground level, bend knees, keep back straight, keep load close to your body, lift with leg muscles, support forearms with knees, and support the load with your body.
- ☆ When lowering a load, use leg muscles and lower the load by bending your knees, not your back. Where possible, support forearms on knees.

Avoid muscle fatigue

- ☆ Warm up first.
- ☆ Take frequent breaks.
- ☆ Change jobs to use different muscles.
- ☆ Gradually get used to the job.
- ☆ Ensure the tractor seat is well sprung.
- ☆ Adopt good posture when standing or sitting at a job.
- ☆ Instead of crouching or squatting for low jobs, use a small stool.

Operating a Tractor

Tractors are the main cause of accidental deaths on farms. Over the years, many farmers, farm workers and others living on or visiting farms, have been killed or seriously injured falling from moving tractors, being run over by tractors, or being crushed when a tractor rolls sideways or backwards.

Regularly check for hazards relating to tractors, attached implements and field conditions. Hazard areas could include mechanical parts, operator training, other people, work procedures, unsafe jacking, climatic conditions, chemicals used, uneven terrain, and any other potential causes of an injury or a hazardous incident. Keep a record to ensure identified hazards are assessed and controlled.

Once a potential hazard has been identified, assess the likelihood of an injury or hazardous incident occurring. For example, risk to children playing near a tractor will vary, depending on what the tractor operator is doing, how close they are to the tractor and whether the operator knows they are there. Consider ways of minimising risk.

Here are some ways of improving tractor operator safety:

- ☆ Read and follow safety procedures in the manufacturer's manual.
- ☆ Ensure an approved cab or roll-over protective structure (ROPS) is fitted.
- ☆ Fit and use a seatbelt on tractors with ROPS.
- ☆ If there is a risk from falling objects, fit a fall-on protective structure (FOPS).
- ☆ To reduce risk of back strain, fit a seat with side restraints and a backrest.
- ☆ Wear hearing protection, and remember, not all tractor cabs are sound proof.
- ☆ Keep children away from tractors and machinery.
- ☆ Remove starter keys when tractors are not in use.
- ☆ Have an up-to-date maintenance schedule.
- ☆ Follow safe maintenance and jacking procedures.
- ☆ Ensure the operator is properly trained for each type of tractor work.
- ☆ Always mount and dismount on a tractor's left side - to avoid controls.
- ☆ Adjust the seat so all controls are safely and comfortably reached.
- ☆ Keep all guards in place, including the power take-off (PTO).
- ☆ Operate the self-starter from the operator position only.
- ☆ Never carry passengers.

When operating a tractor

- ☆ Drive at speeds slow enough to retain control over unexpected events.
- ☆ Reduce speed before turning or applying brakes.
- ☆ Watch out for ditches, logs, rocks, depressions and embankments.
- ☆ On steep slopes, without a trailed implement, reverse up for greater safety.
- ☆ Engage the clutch gently at all times, especially when going uphill or towing.
- ☆ Use as wide a wheel track as possible on hillsides and sloping ground.
- ☆ Descend slopes cautiously in low gear, using the motor as a brake.
- ☆ Never mount or dismount from a moving tractor.
- ☆ Ensure the park brake is on and operating effectively before dismounting.
- ☆ Take short breaks regularly when working long hours.

When towing implements

- ☆ Fit attachments according to the manufacturer's instructions.

- ☆ Always attach implements to the draw bar or the mounting points provided by the manufacturer.
- ☆ Never alter, modify or raise the height of the draw bar unless provided for by the manufacturer.
- ☆ Regularly check safety pins on towed lift-wing implements, to ensure they are not worn.
- ☆ Ensure all guards on towed implements are in place before operating.
- ☆ Never hitch above the centre line of the rear axle, around the axle housing or to the top link pin.
- ☆ Never adjust or work on implements while they are in motion.
- ☆ Never attach implements unless the PTO shaft is guarded.
- ☆ When parking, always lower the three point linkage and towed implement.

To avoid strain injury

- ☆ Adjust the tractor seat for back support and comfort.
- ☆ When buying a tractor, ensure seating is safe and comfortable.
- ☆ Check seat height, seat depth, back rest height and angle, fore and aft movement, seat tilt, firm padding, partial pivoting (if you have to spend long periods looking behind you), and vibration-absorbing suspension.
- ☆ Dismount every hour or so, and spend 5 or 10 minutes doing something active.
- ☆ Plan for your next tractor to include suitably low steps, hand grips, adequate doorway and cab space, and a safe mounting platform.
- ☆ Dismount by climbing down - not jumping down - and use each provided foot and hand hold.

Shearing

Hazards in shearing generally involve machinery, electrical fittings, sheep yard design, slippery and obstructed floors, sharp tools, equipment and protrusions, chemicals, heat stress, and strain injuries from repetitive, awkward and strenuous work.

Conduct a safety audit of shearing sheds, pens, flooring, machinery, wool presses, electrical fittings, connections and cables, lighting, ventilation, and the experience and safety training of those involved, particularly young workers.

Assess identified hazards for likelihood to cause injury or harm. Assess also the potential seriousness of the injury or harm. Consider various safeguards and safe procedures, and assess these for other possible hazards before deciding a plan of action.

Many safety innovations have been developed and implemented to reduce shearing injuries. The following suggestions are to help farmers minimise risks:

- ☆ Design steps, ramps, pens, entrances, flooring, gates and latches to minimise the risk of strain and trip injuries to shearers and helpers.
- ☆ Ensure sheds are well lit and ventilated; cool in summer and draught free in winter.
- ☆ Keep a suitably equipped first aid box in the shearing shed.
- ☆ Have suitable, functional fire-fighting equipment available in shearing sheds and quarters.

Chapter 10

GOOD CUSTOMER INTERACTION

10.1 PROVIDING BASIC CUSTOMER SERVICE

Good customer service relates to the service which you and your employees provide before, during and after a purchase. For example, it's how you interact with your customers. Improving your customer service skills can lead to greater customer satisfaction and a more enjoyable experience for them.

Although it can take extra resources, time and money, good customer service leads to customer satisfaction which can generate positive word-of-mouth for your business, keep your customers happy and encourage them to purchase from your business again. Good customer service can help your business grow and prosper.

Fig Resolving Customers problem

Customer service handles company issues in stores, over the phone and even by email. Companies that produce highly technical products, such as software or satellite television, often have technical support staffs to resolve problems. Customer service is important to an organization for a number of reasons. Foremost, it would be difficult for a company to survive without customer service, as there would be no one available to handle payments or answer questions from prospective customers.

Customer service is important to an organization because it is often the only contact a customer has with a company. Customers are vital to an organization. Some customers spend hundreds and even thousands of dollars per year with a company. Consequently, when they have a question or product issue, they expect a company's customer service department to resolve their issues.

Superior customer service can help your business to grow by increasing:

- ☆ number of customers attracted by favourable word-of-mouth advertising
- ☆ money spent by each customer per transaction
- ☆ frequency and number of times the customers shop

Other benefits of high-quality service include:

- ☆ improved customer satisfaction
- ☆ stronger customer loyalty
- ☆ consistency of work-flow practices
- ☆ reduced marketing costs
- ☆ competitive advantage
- ☆ improved market position
- ☆ staff pride and satisfaction

Customer service is the only way a business can differentiate itself from its competitors. It isn't just about the way you greet and serve your customers. It involves many aspects of business operations, including:

- ☆ products and services information
- ☆ counter and face-to-face service
- ☆ telephone service
- ☆ taking customer orders
- ☆ follow-up documentation
- ☆ billing and managing payments
- ☆ visiting the customer
- ☆ making repairs
- ☆ handling complaints
- ☆ managing the service culture

A company with excellent customer service is more likely to get repeat business from customers. Consequently, the company will benefit with greater sales and profits. Contrarily, companies with poor customer service may lose customers, which will have a negative impact on business. It costs a lot more money for a company to acquire a customer than to retain them, due to advertising costs and the expense of sales calls. Therefore, the efforts that go into maintaining quality customer service can really pay dividends over time.

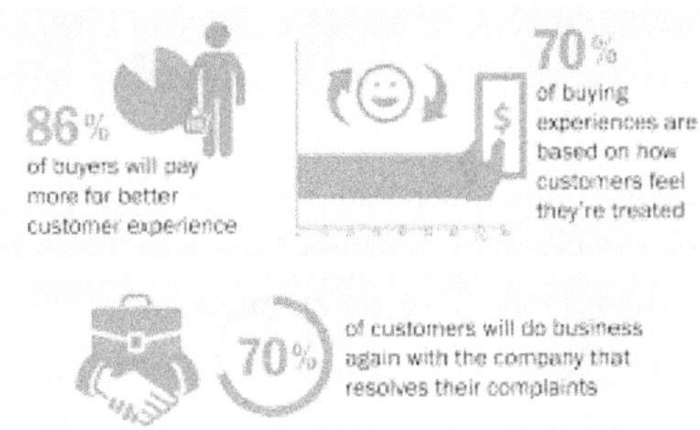

Fig Results of excellent customer service

In your business, customer serve as an important source of information, hence it is required that you should collect required data in order to identify your customers and their behavior which depends on customer to customer profile. If you sell your product to a customer, you will be interested in having their age, gender, income and purpose of buying. In terms of business, you might also eager in knowing the type of industry into which he/she is operating. This will help you in knowing the customer comparability factor in terms of money and type of goods he/she will eager to buy. It is your basic aim to know what your customer thinks about you and your products/ services. This will help in maintaining your database and stock of goods.

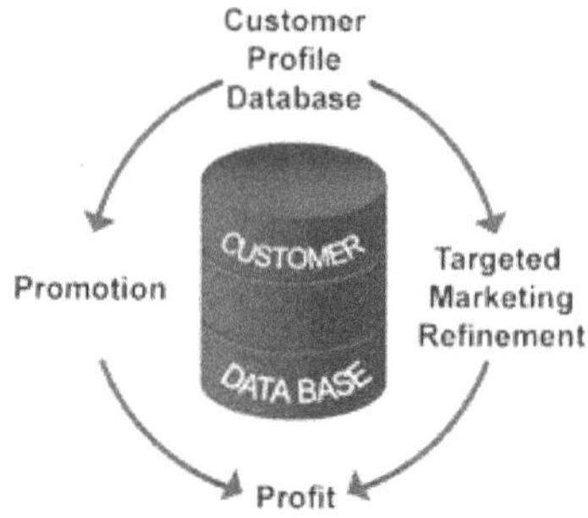

Fig Customers Database

You can ask the customer to give their feed on type of product and services and should also collect information regarding the goods what he/she is searching to buy. Whatever information you collect from the customer should be entered in customer database, so that it will help to fulfill customer's demands in near future.

Having correct information about your customer will help you in maintaining customer profile, from which you can solve many factors related to sale. You should keep the information about your customer related to several questions such as:

- Who they are: This will let you to know about the age and gender of your customer along with its profession.
- What they think and believe: This will let you to have an idea about your customer interests with information or opinion related to you and your product.
- Why, When, Where and how to Pay: This will allow you to know the behavior of your customer related to products they buy, from where they buy, when they buy and which mode they apply for payments.

On collecting information about a customer in above manner will help to sort as per your needs and necessity of your business. While having customise products/ services for various segments will help in focusing the real needs and behavior of your customer. By analysis of your customers will help you to find where that customer will stand in your business. It is noted that certain retail business establishment wants that their customers should always be profitable as possible in:

- buying high margin products
- paying complete price of the product without discounts
- placing small orders of large quantity
- do not cancel or alter their orders
- paying amount on time

☆ not requiring extra after sales service

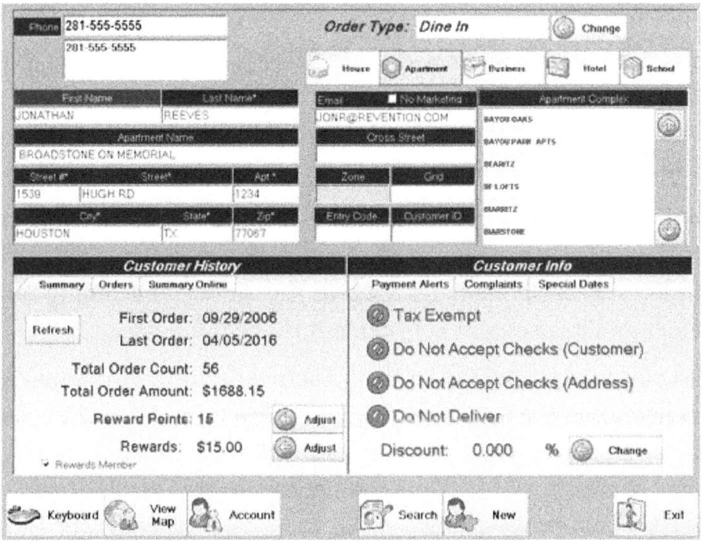

Fig Customer Records

By analysing your customer records you will able to judge their profile and will able to known how profitable each customer would be for you. In some businesses, not many customers are responsible for all your profits as your largest customer might be least profitable.

10.2 GIVING KNOWLEDGE TO CUSTOMER ABOUT THE EQUIPMENT

Product knowledge is the most important tool for closing sales. It instills faith, trust, and respect in the customer, which creates a positive customer experience. Customers have to trust the product, company, and person they interact with before they make a purchase, and by now, you probably know that the person they interact with is weighed much more heavily than the other two sources. This makes it vital for your sales team to be seen as trustworthy sources of information.

10.2.1 Benefits of Product Knowledge

Knowledge is power and for your sales force, product knowledge can be the vehicle to increased sales. We have already discussed the reasons why product knowledge is important to your business, but the list below highlights the benefits of product knowledge—as they directly relate to your sale team.

☆ Strengthen Communication Skills – A thorough and wider understanding of a product enables a salesperson to use different techniques and methods of presenting a product to various types of customers. Stronger communication skills empower a salesperson to suitably adapt a sales presentation for greater impact.

- ☆ Boosts Enthusiasm – Armed with deep product knowledge, a display of enthusiasm and belief in the product may generate excitement among your customers and alleviate uncertainty about the solution that the product provides for the customer.
- ☆ Grows Confidence – If a customer isn't fully committed to completing a sale, the difference may simply be the presence (or lack) of credibility or confidence a salesperson has towards the product. Becoming educated in the product and its uses will help cement that confidence.
- ☆ Assists in Overcoming Objections – Factual information gained from product knowledge, may be used to strike down objections voiced by customers. Solid knowledge about your product coupled with parallel information about similar products sold by your competitors—gives you that added advantage to easily counter objections.

10.3 CONSULTING WITH CUSTOMER RELATED TO PINPOINT PROBLEMS

10.3.1 Customer turnoffs

Many businesses, big or small regularly run into problems of customer service kind. You would think that with abundance of competition, businesses would treat their customers better. Further the common sense isn't all that common and managers and employees alike manage to step on some common customer turnoff mines the moment their customer steps through their door.

Through extensive research customer turnoffs can be categorized into three distinct categories

Value turnoffs

In effect, this means not getting what you pay for. These include inadequate guarantees, a failure to meet quality expectations, and high prices relative to the value perceived. Customers don't expect the same thing from low-cost items as they do from more costly ones, but when characteristics of the core product fail to live up to what customers anticipate, they experience value turnoffs.

Systems turnoffs

These arise from the way a company delivers its products or services. When transactions are unnecessarily complicated, inefficient, or troublesome, customers experience systems turn-offs. Employees who lack the knowledge to answer customer questions, or organizations that have just one person capable of fulfilling a key function, are symptomatic of systems failures. So are telephone menus that are unnecessarily complicated. As per the research, slow service was the main turnoff. But lack of delivery options, cluttered workplaces, unnecessary or repetitious paperwork requirements, poor product selection, and inadequate reordering processes also were mentioned.

People turnoffs

These are the things we most often associate with poor service and include lack of courtesy or attention, inappropriate or unprofessional behavior, or an indifferent attitude -- in short, any behavior that conveys a lack of care or consideration for the customer. In case of service strategy which was in place when customer face lacks of service, in such case, loss of time will make the customer turnoff. That way, we always came out as service winners.

10.3.2. Reasons of Customers complaint

Customers do not complain because they have nothing better to do, but because companies provide less than efficient service, and the quality of their offerings may not be up to the expectations of customers. When customers complain, it usually means that they are trying to highlight a problem with the company's employees, processes, and strategies. Rather than negate the complaint, it would be better for a company to acknowledge the customer's complaint, apologize for the inconvenience, and then swing into action to resolve the problem.

Giving due importance to the words and feelings of customers can save a company from many other problems, and alleviate the possibility of further causes for which customers may complain. Remember, when customers complain, it is an opportunity that a company must be grateful for, and must use it to strengthen the relationship with the customer. Investing in an updated CRM system would enable a company to have every customer's details in a safe and centralized location, easily accessible and retrievable when required. The better control a company can keep on customer information, the greater the possibility of it enhancing service and product quality, which in turn would bring down the number of customer complaints and enhance feelings of warmth and trust towards the company.

When customers complain, the sensible thing to do would be to acknowledge and apologize for the inconvenience, and assure the customer of swift and efficient action, which must be done. Losing even one customer today can prove extremely costly, and companies must first seek to understand the reasons why customers complain, and ensure that they have the capability of resolving the complaints and ensuring they do not reoccur.

There are a number of common complaints that customers make relating to products and services. Having procedures in place that find solutions to common complaints will help you resolve issues quickly and efficiently.

1. Not Keeping Promises

If you give a promise ensure you keep it. Some promises sound great but if they have not been kept this may lead to multiple complaints, distrust and switching to competitors.

2. Poor Customer Service

When customer service representatives are not trained properly, keep customers waiting for a long time and do not resolve issues, this is called poor customer service. It is very irritating to see rep's efforts in resolving your issue and at the end having a simple question unanswered or no solution provided. In addition to this, when customers always hear only your voice menu, see offline live chat button and wait about a week for an email reply, they will definitely start complaining.

3. Transferring from One CSR to Another

Poor staff training leads to calls or chats being transferred from one agent to another. And there is nothing more irritating than telling your story again and again. As you can see a human side of your business can sometimes generate more complaints than your operational side.

4. Rude Behaviour of Staff

Do check conversations and chat records of staff with customers. It'd be a good strategy to keep an eye on this as sometimes rude attitude, inability to listen and interrupting customers may become one of the main reasons for your customers to complain publicly and tell the world about your customer service quality. There are also people who treat customers as if they interfere into their daily routine. The way they communicate with people is enough to turn the most mild mannered of customers against your company. For this:

- ☆ Train staff in client service skills.
- ☆ Develop and use disciplinary procedures.
- ☆ Rotate staff to more suitable duties.
- ☆ Review recruitment and selection procedures.
- ☆ Uninformed staff
- ☆ Make sure staff know your products and services.
- ☆ Make more product literature available for reference.
- ☆ Encourage teamwork and support

5. No Omni-channel Customer Service

Giving your customers an easy way to get in touch with you is one of the first requirements of good customer service. If your customers do not see convenient ways of getting in touch with your team regarding their questions they will contact your competitors and will tell the world about their experience. Today online shoppers use not only email and phone for communication purposes. Live chat and social media are gaining popularity among consumers and for some customers it is convenient to begin conversation using one channel and finish it using another.

6. Not Listening to Customers

Ignoring your customers and not listening to them is one of the biggest mistakes a customer service rep can make and one of the reasons for customers to complain. Just walk a mile in your customers' shoes to see what they feel when they are not listened to.

7. Hidden Information and Costs

Play fair with your customers. Keeping some of your fees hidden and uncovering this in the middle of purchasing process is a frequent reason for a human to complain. The impression given is that your company cheats its customers.

8. Low Quality of Products or Services

If you promise high quality of your products or services and what you deliver is of lower quality, then be sure, your customers will share this bad experience publicly and most likely complain in social media networks.

9. Keeping an issue Unresolved

Unresolved issues and complaints are deadly to any business. If you turn away from your customers and let everything go "as is" then you will lose most part of your customers. But if you keep on resolving issues and complaints improving your service at the same time, then you will win loyal customers who will be glad to share their positive experience.

10. Inaccessibility

How easy is it to contact you? Do you provide enough information about your company? When a customer is seeking assistance there is nothing worse than lack of contact information or unanswered emails and not returned calls. This drives people crazy, especially if they have already paid you.

11. Faulty product

- ☆ Investigate the fault with the manufacturer.
- ☆ Offer a replacement item or replacement model.

- ☆ Repair the product if it is under warranty.
- ☆ Consider replacing the product even if it is no longer under warranty.
- ☆ Train staff to explain warranty conditions at the time of sale and repair.

12. Incorrect descriptions of the product

- ☆ Train staff about the products they are selling.
- ☆ Ensure product literature is accurate.
- ☆ Investigate the product with the manufacturer.
- ☆ Make sure that staff explains warranty conditions.

13. Poor client service

- ☆ Investigate the areas of issue.
- ☆ Train staff in customer service skills.
- ☆ Rotate staff so they can increase their knowledge of other areas.
- ☆ Encourage and support teamwork.

14. Slow delivery

- ☆ Investigate lead times.
- ☆ Re-prioritise delivery schedules.
- ☆ Train staff to provide more accurate delivery data to clients.

15. Poor response to information request

- ☆ Retrain staff in how to respond to client requests.
- ☆ Set procedures on how to respond and response time frames.
- ☆ Provide more take-home literature to customers.

16. Lack of communication

- ☆ Train staff in communication skills.
- ☆ Implement a documentation system (e.g. a hand-over book for shift changes).

10.4 CONSULTING WITH CUSTOMER FOR ANY SPECIFIC NEEDS

Consulting is a process that leads to changes in behaviour and decision-making. It amplifies your clients' abilities to better understand internal processes, strengths, weaknesses, threats and opportunities, and leads to ideas on how to reach the full potential of their company.

Improving discovery requires a process that delivers results quickly without compromising the quality of the insights gained from discovery. Any improvements, therefore, must intentionally focus on how to accelerate existing techniques without necessarily following the exact same procedures to implement them. Three helpful principles to keep in mind:

- ☆ Scale the collection process. Accelerating discovery requires implementation of effective means of scaling existing infrastructure. Technology-enabled solutions are excellent ways to automate previously manual processes. The mantra of automation has overtaken the tech world, and consultants have begun using software to great avail. The key for any consultant is to think of what needs to be scaled in the discovery process and why. As many business leaders are discovering with big data, scaled discovery is not beneficial if it means more noise and less insight. But scaling to gain more meaningful data dramatically improves the quality of discovery.
- ☆ Employ flexible technology. When using technology-enabled solutions, consultants need information that can capture robust data. Since most consultants have their own defined methodologies and frameworks, they need something that can easily translate their conceptual frameworks into useful data without relying on back-end manipulation. Consultants need a way to integrate information into their framework. That way, consultants can easily feed data into their own methodologies without requiring additional work to reconcile what they have with what they need.
- ☆ Make it easy to compare data. Since a consultant's discovery process is about gaining insight from data, it is important to ensure that any technology-enabled solution increases the ease of comparing data. Translating data into a common framework largely services consultants' needs for individual engagements, but firms also need to manage knowledge across multiple engagements. Consultants could add considerable value to their discovery process if they employed technology that used benchmarkable information across other engagements. Not only would benchmarkable data provide a credible baseline to begin discovery, but it would also provide a reasonable model for how to leverage past insights.

Chapter 11

IMPART TRAININGS

11.1 RECRUITMENT AND INDUCTION POLICY FOR CUSTOMER

11.1.1 Recruitment

Every dynamic organization undergoes transitions that require finding, hiring and initiating new employees. Recruitment, selection and induction is the complete cycle of hiring efficient and effective new employees. When management of an entity follows predetermined policies and procedures of recruiting, they make the process transparent, meaning that they comply with affirmative action, are nondiscriminatory and provide equal opportunities to recruits. The processes should be professional and timely. They should comply to certain policies.

A strategic and professional approach to recruitment processes enables the Company to attract and appoint staff with the necessary skills and attributes to fulfill strategic aims and to support Company's value.

The Company is committed to ensure that the recruitment and selection of staff is conducted in a manner that is systematic, efficient and effective, and promotes equality of opportunity.

To provide a flexible framework which promotes good practice, adapts a proactive approach to equality and diversity issues and supports fully the Company's core business whist simultaneously complying with current legislation.

11.1.2 Staff recruitment

The success of the organisation relies on its ability to attract the best staff and volunteers available. Recruitment methods must be fair, efficient, and effective. The Staff Recruitment Policy has been established to ensure organisation has the opportunity to attract the best available staff and volunteers for all vacant positions. This policy relates to employment of all staff and volunteers other than the CEO.

- ☆ The organisation is committed to providing high quality programs and services to our community. To support the achievement of this objective we recognise the importance of employing the most suitable applicant for all vacant positions.
- ☆ The organisation will ensure it has the best opportunity to attract the best available staff by broadly advertising (internally and externally as deemed appropriate) all vacant remunerated positions and volunteer vacancies.
- ☆ The organisation will take all reasonable steps to ensure that applicants may be safely entrusted with the duties of their position.
- ☆ The organisation will internally advertise all vacant positions to current staff and volunteers to encourage career advancement and increase participation.
- ☆ The organisation is committed to providing a work environment that is free from harassment and discrimination.

All recruitment and selection procedures and decisions will reflect the organisation's commitment to providing equal opportunity by assessing all potential candidates according to their skills, knowledge, qualifications and capabilities. No regard will be given to factors such as age, gender, marital status, race, religion, physical impairment or political opinions.

Prior to Conducting the Interview

Review the candidate's resume before commencing the interview. This will help you feel more comfortable when the candidate arrives. Review the similarities or differences in qualifications relating to the performance factors of the job, including:

- ☆ education or basic paper qualifications for the job;
- ☆ related work experience and areas of specialisation;
- ☆ additional experience (such as special interests or volunteer activities) in which the candidate might have developed skills related to the position.

Conducting the Interview

Asking questions is an important part of the interviewer's role; it is not, however, their only responsibility. A good interviewer must also:

- ☆ reduce communication barriers;
- ☆ maintain control of the interview;
- ☆ ensure that the candidate reveals what the interviewer wants to know, not simply what the candidate wants to tell; and
- ☆ create a friendly, conversational atmosphere.

Having the candidate respond to questions and prompts will encourage them to do most of the talking while the interviewer ensures that all relevant topics are covered. The interviewer may be required to ask a question a second time by re-phrasing it or by returning to a particular topic at a later point in the interview. While each interviewer develops a particular style, the following steps provide a useful guide to the structure of an interview.

Step 1: Set the Stage

It's important to create an interviewing environment that allows a candidate to put their best foot forward. An interviewer will be able to gain more information in a comfortable setting and the candidate will be left with a favourable impression of the organisation.

- ☆ Make arrangements for a private meeting room in which to conduct the interview.
- ☆ Do not allow interruptions (e.g. telephone calls etc.).
- ☆ Interviews are more comfortable if conducted in an informal "around the table" setting rather than across a desk, particularly when more than one interviewer is involved. Position the candidate so that they can comfortably direct conversation to anyone in the room.
- ☆ Introduce yourself and all members of the interview panel to the candidate (the panel members may prefer to introduce themselves).
- ☆ Body language should be relaxed and open.
- ☆ Be friendly and courteous throughout the interview. The tone should be like a slightly structured conversation.
- ☆ Sometimes it helps to begin by entering into a general conversation, for example talking about the organization and then asking the applicant to give a summary of their background.

Step 2: Outline the Agenda

Outline for the candidate the structure that the interview will take. This will help them to relax and will put the interviewer in control of what is to follow.

- ☆ Identify areas to be covered (e.g. the duties and responsibilities involved in the job; the candidate's education and experience and how they relate to the position; the use of hypothetical situations).
- ☆ Suggest the length of time that the interview is expected to take, and any additional time that might be spent touring the work site etc.
- ☆ Provide the candidate with a description of the duties and responsibilities of the job and an overview of the workings of the organisation.
- ☆ Avoid confusing or overly technical language. Don't oversell the job or mislead the candidate about the actual duties and responsibilities involved or the future growth expectations of the position.
- ☆ Advise the candidate that there will be an opportunity later in the interview for them to ask questions or add information that may not yet have been covered.

Step 3: Gather Information

Following core questions will provide structure and should take up most of the interview time; however, some flexibility is necessary to allow for follow-up questions and for questions that will arise out of each candidate's documentation. This helps to create a comfortable, relaxed tone.

Listen for evidence of both positive and negative behaviour and focus on one specific performance factor at a time. Analyse how well those behaviours and skills would carry over to the position.

The interviewing process may take some time to master, but it can be extremely effective. Probing is particularly necessary when there are gaps in the candidate's life/work history, when inconsistencies appear or when the candidate changes the subject or is evasive.

Step 4: Welcome Added Information and Answer Questions

In the later stages of the interview, the candidate may have specific questions about the job, department or the organisation itself. A detailed discussion should be reserved until this point, so that the candidate won't simply tailor their answers to suit the position.

11.1.3 Procedures

Pre-Recruitment Activities

- ☆ When it becomes necessary to recruit for a position, Managers should carefully consider the requirements for the position, and the key selection criteria including skills, experience and qualifications.

- ☆ If no position description exists for the available position, or if it requires revising, this is the responsibility of the appropriate Manager. Once the new position description or amendments have been drafted, it should be forwarded on to Human Resources and, if appropriate, approved by the CEO and/or Board. Selection criteria will be drawn up based on the basis of a position statement.
- ☆ Where the position description is for a new role, the Human Resources Officer or Manager will review and evaluate the position and draw up a position statement that will then, if appropriate, approved by the CEO and/or Board.
- ☆ Prior to commencing the recruitment process, the responsible Manager is required to gain approval from the CEO / Board or delegated authority and forward this to the Human Resources Officer.

Direct Internal Appointments/Promotions

In situations where a Manager wishes to promote an employee who meets the specific selection criteria for the vacant position into the internal vacancy, the appointment must be authorised by the appropriate Manager, and the approval is to be forwarded to the Human Resources Department.

Internal Advertising

Where appropriate, the organisation will advertise all vacancies internally. Exceptions to this rule may occur when:

- ☆ The position is of such a specialised nature, and / or appropriate skills are not available within the organisation;
- ☆ There is a need to make a direct appointment or promotion into the vacant position.

Upon receiving approval for the vacant position, Human Resources will advertise the available position internally. Internal advertisements should include the following:

- ☆ Position title;
- ☆ Outline of the position;
- ☆ Skills required for the role;
- ☆ Closing date for applications.

All internal applicants should forward a current copy of their resume, together with covering letter, to the applicable manager for acknowledgement, consideration and processing.

Internal applicants who possess the required skills, qualifications and work-related experience, as specified in the internal advertisement, will be interviewed for the position by the relevant Manager or Supervisor.

External Advertising

Where a position cannot be filled internally or where it is appropriate to conduct an external recruitment campaign, the available position should be advertised through relevant networks, on relevant websites, and through local employment services.

Volunteer positions will be advertised as widely as deemed reasonable.

All advertisements must be approved by the CEO.

If required, the Human Resources Department will prepare an appropriate recruitment advertisement for the position and submit it for review and approval by the relevant Manager. The Human Resources Department will administer the placement of the advertisement and monitor applications received.

Use of Recruitment Consultants

Where deemed appropriate, external recruitment consultants may be used for recruitment purposes. The Recruiting Manager should contact the Human Resources Department for assistance in engaging the services of recruitment consultant.

It remains the relevant Manager's responsibility to ensure that the recruitment consultant adheres to [Name of Organisation]'s recruitment and selection policies.

Screening Applicants

- If a recruitment consultant has been engaged to recruit for a position, they will be responsible for screening the applicants.
- Resumes must be screened against the position description so that assessments can be made of their suitability for the specific role. Applicants who are assessed as suitable will then be selected for interview.
- Managers should consult with the Human Resources Department if they require any assistance with the selection process.
- Where appropriate, but particularly in positions of financial responsibility or in dealing with vulnerable clients / children, police checks may be arranged. Police checks shall be arranged only with the consent of the applicant concerned; however, if consent is refused this shall be taken into consideration in the selection process.
- References shall be sought, where appropriate, as set out in the organisation's References Policy. Previous employers and referees shall be contacted, and transcripts, qualifications, publications and other certification or documentation shall be validated.
- Any checks which may form part of the selection process should be conducted prior to issuing an offer of employment.

Conducting Interviews

- ☆ The short-listing and interview process will be conducted by a selection panel which will be appointed by and will include the CEO or their nominee and the relevant manager or supervisor for the position.
- ☆ If any member of staff finds that they are assessing any applicant where there is a perceived or actual conflict (eg. Where the applicant is a family member, friend or past colleague) they shall declare the perceived or actual conflict to the panel.

Reference Checking

- ☆ Managers are to ensure that, where possible, a minimum of two reference checks are conducted prior to an offer of employment being extended to a candidate.
- ☆ Details of the reference checks should be attached to the candidate's application for future reference.

New Starter Paperwork

- ☆ If an internal candidate is selected, the Manager is required to notify the successful candidate and their Manager. If an external candidate has been selected, the Manager is to make a verbal offer to the candidate.
- ☆ To authorise the commencement or transfer of an internal employee, the Manager must notify the Human Resources Department and provide confirmation of the CEO's approval. The Manager should ensure that all recruiting documents are completed and returned to the Human Resources Department for filing.
- ☆ The Human Resources Department will prepare a written letter of offer for the successful candidate. The letter of offer and or contract of employment will confirm the start date, salary (if any), position and the terms and conditions of employment pertaining to the employee.
- ☆ Once the Human Resources Department or Manager has received the candidate's signed letter of offer, the Human Resources Department is to notify all unsuccessful candidates. If an external recruitment agency has been used, the Manager is to notify the agency, who will notify the unsuccessful candidates.
- ☆ The Manager is responsible for liaising with the Human Resources Department to ensure that the necessary documentation, equipment and access privileges are prepared for the new employee.
- ☆ The Human Resources Department will forward an induction kit to the new employee for their completion.

Records and Correspondence

- ☆ All contact regarding the position is to be directed through reception, with all applications marked "Confidential" and posted to the CEO.
- ☆ Letters/emails of acknowledgment should be posted to all applicants prior to the short-listing of final suitable applicants. Short-listed but unsuccessful applicants should be advised that their CV will be retained by the Human Resources Department for future reference, unless the applicant advises otherwise.
- ☆ Applicants who do not meet the key selection criteria and are not suitable to be short-listed for an interview should be sent a written letter advising them that their application has been unsuccessful.

11.1.4 Induction Policy

1. General

The company believes that all new employees must be given timely induction training. This training is regarded as a vital part of staff recruitment and integration into the Robinson working environment. The policies, associated procedures and guidelines define the Company's commitment to ensure that all staff is supported during the period of induction to the benefit of the employee and Company alike.

2. Aim

It is the aim of the company to ensure that staff induction is dealt with in an organised consistent manner to enable staff to be introduced into a new position and working environment expediently, so that they can contribute effectively. The implementation of good induction practices will ensure:

- ☆ Enabling new employees to settle into the Company and become productive efficient members of staff within a short period of time.
- ☆ Ensure that new entrants are highly motivated and that this motivation is encouraged.
- ☆ Assist in reducing staff turnover, lateness, absenteeism and poor performance.
- ☆ Assist in developing a management style where emphasis is on leadership.
- ☆ Ensure that employees operate in a safe working environment.
- ☆ Will reduce costs associated with repeated recruitment, training, lost productiveness and time. Ensuring of consistent level of service.

3. The Company's Commitment

The company will:-

- ☆ Issue guidelines to familiarise staff with the induction process.
- ☆ Ensure all staff is issued with a full Induction Pack.
- ☆ Maintain and update Induction Policies
- ☆ Ensure there is effective monitoring of the induction process particularly in the first three months.
- ☆ Deals with any problems that promptly provides an efficient service for both Managers and Staff.
- ☆ Review all policy, procedure and guideline documents on a regular basis.
- ☆ Provide relevant formal training courses necessary to assist the induction process.

4. Completing the Induction Process

This depends on each individuals need. For some staff, induction may be completed in a number of days. For others where the role is complex or specific training needs have to be addressed, induction may take several weeks. It is a matter of balancing being both efficient and effective. It further ensures that new employee is given the opportunity to contribute to the overall improvement and constant review of all our methods and practices.

General organisational induction training checklist:

- ☆ Essential 'visitor level' safety and emergency procedures
- ☆ Washrooms
- ☆ Food and drink
- ☆ Smoking areas and policy
- ☆ Timings and induction training overview
- ☆ Organisational history and background overview
- ☆ Ethics and philosophy
- ☆ Mission statement(s)
- ☆ Organisation overview and structure
- ☆ Local structure if applicable
- ☆ Departmental structure and interfaces
- ☆ Who's who (names, roles, responsibilities)

- ☆ Site layout
- ☆ Other sites and locations
- ☆ Dress codes
- ☆ Basic communications overview
- ☆ Facilities and amenities
- ☆ Pay
- ☆ Absenteeism and lateness
- ☆ Holidays
- ☆ Sickness
- ☆ Health insurance
- ☆ Pension
- ☆ Trades Unions
- ☆ Rights and legal issues
- ☆ Personnel systems and records overview
- ☆ Access to personal data
- ☆ Time and attendance system
- ☆ Security
- ☆ Transport and parking
- ☆ Creche and childcare
- ☆ Grievance procedures
- ☆ Discipline procedures
- ☆ Career paths
- ☆ Training and development
- ☆ Learning Styles Self-Assessment
- ☆ Multiple Intelligences Self-Assessment
- ☆ Appraisals
- ☆ Mentoring
- ☆ Awards and Incentives
- ☆ Health and Safety, and hazard reporting
- ☆ Physical examinations, eye test etc.

- ☆ Emergency procedures, fire drill, first aid
- ☆ Accident reporting
- ☆ Personal Protective Equipment
- ☆ Use, care, and issue of tools and equipment
- ☆ Other housekeeping issues
- ☆ General administration
- ☆ Restricted areas, access, passes

Job and departmental induction training checklist

The induction training process also offers the best opportunity to help the new person more quickly integrate into the work environment - particularly to become known among other staff members. Hence the departmental tours and personal introductions are an absolutely vital part of induction. Organisations depend on its people being able to work together, to liaise and cooperate - these capabilities in turn depend on contacts and relationships.

- ☆ Local departmental amenities, catering, washrooms, etc.
- ☆ Local security, time and attendance, sickness, absenteeism, holidays, etc.
- ☆ Local emergency procedures
- ☆ Local departmental structure
- ☆ Department tour
- ☆ Departmental functions and aims
- ☆ Team and management
- ☆ People and personalities overview (extremely helpful, but be careful to avoid sensitive or judgemental issues)
- ☆ Related departments and functions
- ☆ How the department actually works and relates to others
- ☆ Politics, protocols, unwritten rules (extremely helpful, but be careful to avoid sensitive or judgemental issues)
- ☆ The work-flow - what are we actually here to do?
- ☆ Customer service standards and service flow
- ☆ How the job role fits into the service or production process
- ☆ Reporting, communications and management structures
- ☆ Terminology, jargon, glossary, definitions of local terms

- ☆ Use and care of issued equipment
- ☆ Work space or workstation
- ☆ Local housekeeping
- ☆ Stationery and supplies
- ☆ Job description - duties, authority, scope, area/coverage/territory
- ☆ Expectations, standards, current priorities
- ☆ Use of job specific equipment, tools, etc.
- ☆ Use of job specific materials, substances, consumables
- ☆ Handling and storage
- ☆ Technical training - sub-categories as appropriate
- ☆ Product training - sub-categories as appropriate
- ☆ Services training - sub-categories as appropriate
- ☆ Job specific health and safety training
- ☆ Job-specific administration, processing, etc.
- ☆ Performance reporting
- ☆ Performance evaluation
- ☆ Training needs analysis method and next steps
- ☆ Initial training plans after induction
- ☆ Training support, assistance, mentor support
- ☆ Where to go, who to call, who to ask for help and advice
- ☆ Start of one-to-one coaching
- ☆ Training review times and dates
- ☆ Development of personal objectives and goals
- ☆ Opportunities for self-driven development
- ☆ Virtual teams, groups, projects open to job role
- ☆ Social activities and clubs, etc.
- ☆ Initial induction de-brief and feedback
- ☆ Confirmation of next training actions
- ☆ Wider site and amenities tour

11.2 TRAINING MATRIX

This matrix has been to help identify where specific training needs exist and may be completed by the front line employees who have the customer contact. It is based on the relationship of particular levels of importance to the current role and the perceived skill/standard required of the individual person. When you develop a training matrix you need to get a list of the employees that will be included in the matrix. Organize the employees into to groups that are similar to the organizations structure and list the required skills needed for each job role.

Staff should work with supervisors to select training that bests fits professional growth and job requirement needs. Employees may use additional in-service hours over and above the District Directed hours required to attend below training if hours are available. Priority should be given to training that is required by the District and/ or for your position before encouraged training. The following are example modules for Achieving Excellence in Customer Service training for front-line workers and first-line supervisors.

Module 1 – Compassion for the Customer

- ☆ As a result of this training module, participants will be better able to:
- ☆ Identify customers and what they value.
- ☆ Identify and distinguish Moments of Truth, Moments of Misery, and Moments of Magic for the customers.
- ☆ Describe the most difficult customer ever encountered and provide a theory as to why the customer behaved as s/he did.
- ☆ Explain why the customer is the job, rather than an interruption of it.
- ☆ Describe the 6 basic customer needs, knowing that customers ask for what they want, but what they need is a whole lot more.
- ☆ Explain ways in which co-workers are internal customers and why they should be treated with the same respect as external customers.
- ☆ Describe 4 basic customer working styles and how to get in step with these working styles.
- ☆ Identify 10 personal skills that influence other people's behavior positively.

Module 2 – Listening and Problem-Solving Skills

- ☆ As a result of this training module, participants will be better able to:
- ☆ Demonstrate successful attending skills for listening, namely:
- ☆ Presence

- ☆ Attention
- ☆ Authentic Interest
- ☆ Suspension of Judgment
- ☆ Patience
- ☆ Commitment to Work
- ☆ Self-identify tendencies that block attending, namely:
- ☆ Ruling out the speaker
- ☆ Reaching a premature conclusion
- ☆ Reading in expectations
- ☆ Reading out threats
- ☆ Rehearsing a response
- ☆ Reacting to trigger words
- ☆ Responding with evaluation
- ☆ Differentiate between open and closed questions and use them appropriately to serve your customers better.
- ☆ Explain and demonstrate how asking the right questions can help identify the three different kinds of customer-based opportunities
- ☆ Service Opportunities
- ☆ Persuasion Opportunities
- ☆ Tension Opportunities
- ☆ Identify and demonstrate the steps in solving a problem for a customer, including the questioning technique and the bridging technique.

Module 3 – Communication and Telephone Skills

- ☆ As a result of this training module, participants will be better able to:
- ☆ Use inflection, pacing, and volume to achieve a great telephone voice.
- ☆ Demonstrate proper telephone etiquette for how to answer the phone and for putting callers on hold.
- ☆ Use behavioral flexibility in communicating successfully with people who have different working styles.
- ☆ Check the bonds of trust and communications accuracy regularly in a two-way conversation.

Module 4 – Handling Tense, Stressful and Multi-Tasking Situations

- ☆ As a result of this training module, participants will be better able to:
- ☆ Explain the concept of Tension Management
- ☆ Practice application of Tension Management to common situations that you experience.
- ☆ Describe and demonstrate the best approach for handling disgruntled, unhappy, or irate customers.
- ☆ Let customers vent their feelings without taking personal offense.
- ☆ Avoid negative filters about the customer which ignite additional tension.
- ☆ Reach mutual agreement on the solution with the customer so that you under-promise and over-deliver.

Chapter 12

INTRODUCTION TO SAFETY, HEALTH AND ENVIRONMENT

12.1 HEALTH AND SAFETY TRAINING

1. Eye protection shall be worn at all times when operating power equipment and tools.
2. Report all equipment defects immediately. Do not use unsafe equipment until it is repaired.
3. When disconnecting equipment, pull on the plug, not the cord. Whenever the use of electrical extension cords is required, keep them clean, dry, and free from kinks. Protect them from oil, hot or sharp surfaces, and chemicals. Exterior cords should not be extended across aisles, through water, doors, or into areas where they are apt to be damaged.
4. Use equipment properly; do not overload the motor by forcing cutting tools into the work.
5. Do not leave power tools unattended while in operation.
6. Portable electric circular and band saws shall not be used unless the guard mechanism is functioning properly.
7. Oversized drill bits shall not be ground to fit smaller electric drills.
8. Long hair and loose clothes must be restrained in a manner that will not allow them to be caught in machines.

12.2 SAFETY RULES RELATED TO DIFFERENT MACHINES

12.2.1 Grinding Machines

1. Grinding wheels and wire brushes shall not be operated in excess of the speed recommended by the manufacturer. Check the recommended rpm against that of the shaft or motor before mounting a new wheel. Check all grinding wheels for chips and cracks before use.
2. Face shields, safety glasses, or chipper's goggles shall be worn at all times when grinding or using a wire brush.
3. Gloves shall not be worn while grinding, nor will cloth be used to hold work pieces.
4. Do not operate grinding machines unless metal wheel hoods are in place. Do not apply work too quickly to a cold wheel.
5. Tool rests shall be secured at all times and adjusted to within 1/8 inch of the wheel. Top wheel guards shall be adjusted to within 1/8 inch of the top of the wheel.
6. Disc grinder tables shall be adjusted to within 1/8 inch of the disc.
7. When a grinder is first turned on, do not stand in line with the grinding wheel. If any wobble or vibration is noticed, the machine must be turned off and repaired.
8. Except where specifically designed, one should not grind on the sides of the grinding wheel.
9. Do not grind wood, aluminum, copper, or other soft materials on wheels designed for steel and iron.

12.2.2 Metal Lathes

1. Chip guards should be used in operations that could endanger the operator or others nearby.
2. Chip breakers shall be used whenever practical. Tool ways must be kept clear and clean.
3. Tools should not be set or adjusted while the lathe is in operation. Tools and chucks must be checked for defects before use.
4. Brushes or chip pullers shall be used for removing chips. Operators shall not use their hands, or compressed air in excess of 30 psi, to remove chips.
5. Heavy chucks, face plates, or other heavy equipment should never be handled without proper lifting equipment.
6. Tools or other equipment shall not be stored on top of the head stock.

7. Rotating stock extending into aisles should be marked with a warning device (yellow tape, rag, tag) or contained by physical barrier.
8. Magnesium or similar metals shall not be machined unless appropriate fire protection is provided.
9. Do not stop lathe with tool bit in the cut, or with feed clutch engaged. Hand pressure should not be used to stop free spinning chucks.

12.2.3 Metal Cutting Band Saws

1. Before starting an operation, be sure the machine is set for the recommended speed, feed, and blade type for the material to be cut.
2. A complete face shield shall be worn when blades are electrically welded on the machine.
3. The portion of the blade between the upper wheel and the saw table should be completely enclosed except for the point at which the cut is made.
4. Inspect and adjust the table and blade guide to be sure that small parts cannot jam between the table and moving blade.
5. The length of the exposed blade should not be more than 3/8 inch greater than the thickness of the stock being cut.
6. Use pliers, tongs, jigs, or other hold-down devices when sawing small parts that could jam between the blade and saw guide.
7. Warn personnel or install barriers during sawing or welding operations that throw hot sparks onto nearby work stations.

12.2.4 Metal Planers, Shapers, drilling and Boring Machines

1. Always use brushes or chip pullers to clean the work area. Operators shall not use their hands or compressed air in excess of 30 psi to remove chips.
2. Always clamp the work securely before starting the cut. Do not measure the job while the machine is in operation.
3. Always remove the stroke-change screw handle before starting the shaper.
4. Do not place heavy parts or tools on the machine without the use of approved lifting equipment.
5. Only soft metal or plastic hammers should be used when setting up jobs on a drill press or boring mill.
6. Adjustable wrenches should not be used on the machine parts or equipment. Properly sized box or open-end wrenches should be used.
7. Do not operate drill presses with dull tools.
8. Never make adjustments on the chuck when the machine is in motion.

9. Boring mill operators should never attempt to make measurements near the tool, reach across the table, or adjust the work while the spindle is turning.
10. When deep holes are being drilled beyond the flutes of the drill, the drill should be withdrawn frequently to keep it free of chips.
11. Stop the machine before attempting to clear work that has been jammed.

12.2.5 Power Presses and Forming Equipment

1. Do not operate power punch presses without "point of operation guarding." Do not remove or modify guards.
2. Power presses shall not be operated in the continuous tripping cycle unless the point of operation is guarded on all sides by approved barrier guards.
3. Safety tongs shall be used whenever it is necessary to reach into the point of operation of any machine.

12.2.6 Explosive Actuated Tools

1. Explosive actuated tool operators must be trained and certified for this work.
2. Always wear safety goggles to avoid the possibility of flying chips, etc.
3. Never, under any conditions, attempt to discharge a stud or pin into free flight.
4. Use only the appropriate boosters, studs, and pins designed for the tool.
5. Never drive a stud or pin into extremely hard surfaces such as glazed tile, glazed brick, glass, tool steel, etc. Such surfaces may cause a ricochet.
6. Do not drive fasteners closer than 3 inches from the edge of concrete, brick, or other like materials.
7. Do not drive fasteners closer than 1/2 inch from the edge of steel.
8. In the event of misfire the tool is to be held in operating position for a minimum of 30 seconds before disassembly.

12.2.7 Handling Chemicals

1. Do not wash hands in cleaning solvents. Absorption of these liquids through the skin can cause serious illness.
2. Do not handle chemicals of any type unless you are aware of the potential skin and inhalation hazards. Consult the appropriate MSDS.
3. Wear appropriate skin, face, eye, and hand protection when moving or handling bulk chemicals.
4. Always wear chemical resistant gloves, aprons, and complete eye protection when handling corrosive chemicals. If chemicals contact skin, wash with large amounts of water immediately.

5. Do not open chemical containers that have been stored in the sun unless proper care is taken. Many chemicals will build up pressure in the container when exposed to heat.
6. Clean up small chemical spills immediately if you can do it safely; otherwise, notify supervisor.

12.2.8 Forklift Operations

1. Only authorized, trained personnel shall operate lift trucks.
2. Before start of shift, a visual inspection must be made to ensure that horn, lights, brakes, tires, gas supply, hydraulic lines, etc. are in safe working condition. Employees shall not operate an unsafe forklift at any time.
3. Fill fuel tanks out of doors while engine is off.
4. Do not exceed the safe load capacity of a forklift at any time. Do not counterweight a forklift to increase lifting capacity.
5. Operators shall drive with both hands on the steering wheel. Horseplay is prohibited. Do not drive with wet or greasy hands.
6. No person shall ride as a passenger on a forklift or on the load being carried.
7. A forklift will not be used to elevate a platform or pallet with persons on it, except work platforms especially designed for this purpose. Work platforms must have standard guard rails, and must be securely fastened to the forks.
8. No person shall stand or walk under elevated forks.
9. Operators should avoid making jerky starts, quick turns, or sudden stops. The operator will not use reverse as a brake.
10. Forklifts should be driven on the right side of the road or aisle-way.
11. Operators shall cross railroad tracks diagonally whenever possible.
12. Forklifts shall be operated at a safe speed with due regard for traffic and conditions. Maximum speed limits: inside buildings, 5 mph; outside buildings in work areas, 7 mph; on roads, 10 mph.
13. Slow down on wet and slippery surfaces and at cross aisles or locations where vision is obstructed.
14. Operators entering a building or nearing a blind corner shall make their approach at reduced speed. Sound horn and proceed carefully.
15. Standard arm signals will be used at all times.
16. Operators shall give pedestrians the right-of-way at all times.
17. Operators shall not drive toward any person who is in front of a fixed object or wall.
18. Operators shall not overtake and pass another forklift traveling in the same direction, at intersections, blind spots, or hazardous locations.

19. Operators should not put their fingers, arms, or legs between the uprights of the mast, or beyond the contour of the forklift.
20. When the forklift is not carrying a load, the operator shall travel with the forks as low as possible (maximum of 3 inches on paved surfaces). When carrying a load, it should be carried as low as possible (consistent with safe operation, 2 to 6 inches above the surface.)
21. Forks should always be placed under the load as far as possible. Do not lift a load with one fork.
22. No load should be moved unless it is absolutely safe and secure.
23. The operator's view should not be obstructed by the load. In the event of a high load, the forklift will be driven backward.
24. Operators shall look in the direction of travel.
25. The forks should not be operated while the forklift is traveling.
26. On a downgrade, the load shall be last, and the forks raised only enough to clear the surface.
27. On an upgrade, the load shall be first, and the forks raised only enough to clear the surface.
28. Use extra care when handling long lengths of bar stock, pipe, or other materials.
29. Avoid sharp or fast end-swing.
30. Compressed gas cylinders shall be moved only in special pallets designed for this purpose.
31. When unloading trucks or trailers, the brakes on the vehicle will be set (locked) and the wheels chocked.
32. Forklifts must be safely parked when not in use. The controls shall be neutralized, power shut off, brakes set, key removed, and the forks left in a down position flat on the surface, and not obstructing walkways or aisles.
33. A forklift shall not be left on an incline unless it is safely parked and the wheels blocked.

12.3 PRECAUTIONS AT WORKPLACE

1. **Wear approved eye protection.** Industrial-quality eye protection should be worn at all times. Safety glasses should have the Z87.1 logo on them to assure they are industrial quality.
2. **Wear proper clothing.** This varies depending on the type of hand tool you are working with. Work clothing should not be loose, baggy, or highly flammable. To protect against burns, wear clothing such as coveralls, high-top shoes, leather aprons and leather gloves. Remove all paper from pockets

and wear cuffless pants. When working with heavy metals wear hard-toed shoes with non-skid soles. Avoid wearing synthetic clothing because it has a low flashpoint, which can result in severe burns. Do not wear jewelry. It can get caught in moving parts.

3. **Protect your hair, scalp, and head.** Pull back long hair in a band or a cap to keep it from getting caught in tools. Be extremely careful with long hair when using a drill or drill press. When handling carpentry materials wear a hard hat or bump cap to protect your head.
4. **Watch your fingers.** Take special care when hammering so that you strike the object, not your fingers.
5. **Keep your mind on your work.** Avoid horseplay and loud talk. Loud talking as well as pushing, running, and scuffling while working with hand tools can cause serious accidents. Be alert and work defensively.
6. **Keep work area and tools clean.** Dirty, greasy, and oily tools and floors can cause accidents. Clean and put away all unneeded tools and materials. Clean up spills and scraps from the floor and equipment. Keep paths to exits clear. If conditions are dusty, use a respirator.
7. **Use tools properly.** Always use proper-sized tools and equipment for the job. Use each tool only for the job for which it was intended. Forcing a small tool to do the job of a large one may result in injury or tool damage. Never use a screwdriver to see if electrical circuits are hot. Never use a machinist›s hammer in place of a carpenter›s hammer. Do not strike a hardened steel surface, such as an anvil, with a steel hammer because a small piece of steel may break off and injure someone. Be sure wrenches fit properly. Never use pliers in place of a wrench. Never strike wrenches with hammers. Pull on wrenches, do not push. When sawing secure the material in the saw vise.
8. **Keep cutting-edge tools sharp.** Dull cutting-edge tools are dangerous, as they require excessive pressure and hammering to make them cut. When cutting always cut away from the body. Before using any cutting tool, remove nails or other objects that might destroy the tool›s cutting edge.
9. **Carry and store tools properly.** All sharp-edge tools and chisels should be carried with the cutting edge down. Never carry sharp tools in a pocket. Store all sharp-edge cutting tools with the sharp edges down.
10. **Inspect tools before using.** Avoid using damaged tools. Tools that appear to be damaged or have broken handles should be marked unsafe. Do not use them until they have been repaired.
11. **Grip tools firmly.** Hold hand tools securely so that they do not slip and hit someone. Do not wear gloves--they are bulky and make gripping tools difficult.

12.4 VEHICLE MAINTENANCE SAFETY RULES

1. Eye protection is mandatory for all operations which produce sparks, chips, flying objects or involve use of corrosive chemicals. Face shields shall be worn for all operations that involve use of a high-pressure steam system. Appropriate gloves and protective clothing shall also be worn.
2. Mechanics shall not wear loose clothing around rotating equipment. Clothes saturated with oil, grease, or solvents shall not be worn.
3. Compressed air shall not be used to clean clothing.
4. Shop floors will be kept free of grease, oil, gasoline, or other slipping hazards.
5. Employees shall not use defective electrical or mechanical shop equipment or hand tools. All automotive shop machinery shall be grounded.
6. Vehicles shall not be towed unless appropriate tow bars or other approved equipment is used.
7. Jacks, hoists, or other lifting devices shall not be used beyond the safe load capacity recommended by the manufacturer. Employees shall not remain in vehicles being lifted by hydraulic lifts or jacks.
8. Mechanics shall not work under vehicles that are not properly supported with approved stands. Makeshift stands made of wood, cement blocks, or boxes shall not be used.
9. Gasoline, acetone, kerosene, or similar solvents shall not be used to clean hands, floors, walls, or other surfaces. Parts shall be cleaned only in approved containers using appropriate solvents.
10. Employees shall not use standard sanitary sewer drains for the disposal of gasoline, oil, or solvents. Contact EH&S for disposal guidelines.
11. Tanks or containers that are used for gasoline or other flammable solvents shall not be mechanically opened or repaired by welding without purging and cleaning.
12. Do not begin tire inflation before the rim is properly seated. It is dangerous to attempt adjustment with a hammer when the tire is being inflated.
13. Do not place hands or arms between mounted dual tires during inflation. Always use a long air chuck for inflation.
14. Do not change tires on the road unless wheel chocks and warning devices are used. Flares should be used to warn others whenever a vehicle tire is changed while on a heavily used road.
15. Changing of tires on split-rim wheels will be performed only by individuals with proper training and using only appropriate equipment.

Chapter 13

TYPES OF AGRICULTURAL HAZARDS

13.1 MACHINERY HAZARDS

There are a multitude of machines used in mechanized agriculture. These machines are powered in many different ways including PTO shafts, hydraulic oil pressure, electrical power, engine power and ground traction. Many machines have several types of hazards. Table gives machine hazards, descriptions of the hazards and examples of where the hazards occur on various machines.

Table Common machinery hazards and where they occur

Hazards	Sources	Locations
Pinch points	Two machine parts moving together with at least one of them moving in a circle	Where drive belts contact pulley wheels, drive chains contact gear sprockets, feed rolls mesh together
Wrap points	An exposed/unguarded rotating machine component	Power take-off (PTO) drive shafts, beater bars on self-unloading ensilage wagons, blades of some manure spreaders
Shear/cutting points	The edges of two moving parts move across one another, or a single edge moves against a stationary edge or soft material	Mowers and forage harvesters, small-grain combine heads, bedding choppers, grain augers

Hazards	Sources	Locations
Crush points	Two moving objects moving toward each other, or one moving object moves toward a stationary object	The front and rear tires/sections of articulating tractors, hitching machinery, a hand caught under a piece of hydraulically-controlled equipment
Free-wheeling parts	Machine parts that continue to move after power to the part has stopped, usually from the continuing rotation of knife or fan blades	Forage harvesters, feed grinders, rotary mowers, ensilage blowers
Thrown objects	The chopping, grinding, cutting, and flinging motions of machines. Small objects such as rocks, metal, glass, sticks, and vegetation may be picked up and thrown with great force	Rotary mowers, feed grinders, combines with straw choppers, and manure spreaders
Stored energy	Energy that is confined and released unintentionally or unexpectedly	Machine springs, hydraulic systems, compressed air, electrical systems
Burn points	Skin burns from contacting hot parts of machines	Hot mufflers, engine blocks, pipes, fluids (fuel, oils, chemicals)
Pull-in points	Occurs at the point where the machine takes the crop material in for further processing	Corn pickers and combines, forage choppers, and hay balers
Noise-induced hearing loss	Operating machinery	Tractors, field machinery, grain augers, dryers, silo blowers, bedding choppers, feed grinders. Damaging noise level may come from a combination of one or more machines. Older machines generally produce louder sounds than newer machines.

13.1.1 Machinery power and speed

Though workers may understand that machinery is powerful and operates at very high speeds, most workers have not stopped to consider just how powerful machines are in comparison to their own power, nor do they fully comprehend how fast machines are. Machinery power varies considerably, but even small machines generate many times more horsepower than any person. A quick, pull-away action of a human arm normally generates less than 1 horsepower (hp), sometimes much less. A small 16-hp machine, such as a walk-behind mower, may have 20 to 40 times more power pulling a person into the machine than that person can generate pulling away. A medium-sized machine operated at 40 to 60 hp will have hundreds of times more power than a person.

This power and speed combination presents many potentially hazardous situations to workers. For example, the tractor's PTO stub shaft transfers power between the tractor and PTO-powered machinery. Power transfer is accomplished by connecting a drive shaft from the machinery to the tractor's PTO stub. The PTO stub and drive shaft rotate at 540 rpm (9 times/second) or 1,000 rpm (16.7 times/second) when operating at full recommended speed. Most incidents involving PTOs stem from clothing suddenly caught by an engaged but unguarded PTO stub or driveline. Even with a relatively quick reaction of 1 second (i.e., the worker tries to pull away from the shaft) and a shaft with a diameter of 76 mm operating only at half speed (e.g., at 270 rpm (one-half of 540), the victim's clothing has already wrapped 1.1 m around the shaft. A faster-operating PTO and/or a slower reaction provides even less of an opportunity for the worker to avoid entanglement with the shaft.

When a machine is running at full recommended PTO speed, crop material moves into the machine intake or processing area at approximately 3.7 m/s. If a worker is holding onto crop material as it begins entry into the machine, he or she is usually unable to let go quickly enough to release the material before being pulled into the machine. In 0.3 second, the worker will be pulled 1.1 m into the machine. This situation most often happens when crop material plugs the intake point of the machine and the worker attempts to unplug it with the PTO engaged.

13.1.2 Machinery safety

Machinery safety is largely a matter of keeping the guards and shields that came with the original in place and properly maintained. Warning decals should be used as a reminder to keep guards and shields in place. If guards or shields must be removed for maintenance, service or adjustment, they must be replaced immediately upon completion of the repair. Safe operating practices must be followed. For example, the tractor must be shut off and the PTO or block hydraulic systems disengaged before unplugging or servicing equipment. Operator manuals must be read and their safety instructions followed. Workers must be properly trained.

13.2 HAZARDS DUE TO OPERATING OF TRACTOR

Tractors are the main cause of accidental deaths on farms. Over the years, many farmers, farm workers and others living on or visiting farms, have been killed or seriously injured falling from moving tractors, being run over by tractors, or being crushed when a tractor rolls sideways or backwards.

Regularly check for hazards relating to tractors, attached implements and field conditions. Hazard areas could include mechanical parts, operator training, other people, work procedures, unsafe jacking, climatic conditions, chemicals used, uneven terrain, and any other potential causes of an injury or a hazardous incident. Keep a record to ensure identified hazards are assessed and controlled.

Farm tractors have many characteristics that result in their being the most important piece of power equipment on the farm. Most tractors have rubber tyres, hydraulic systems, and power take-off (PTO), and utilize a combination of engine speeds and gear ratios. These characteristics combine to provide tractors with speed, power, flexibility and adaptability. The most serious hazards associated with tractor operation include overturns, runovers and PTO entanglement. Tractor overturns fatally injure far more victims than any other type of incident. Table provides a listing of tractor hazards and how injuries occur.

Table Common tractor hazards and how they occur

Hazard	Type of incident	How injury occurs
Overturns	Side rollovers	Operating on slopes, turning corners too fast, rear wheel drops into a hole or off-road surface.
	Rear rollovers	Hitching to a point other than the drawbar, rear wheels are stuck in mudhole or are frozen to the ground.
Runovers	Passenger (extra rider) falls off	Most tractors are designed only for one operator; therefore, there is no safe location for an extra person on a tractor.
	Operator falls off	Knocked off by low-hanging tree limb, bounced out of seat by traversing rough ground.
	Operator is run over while standing on the ground	Jump starting tractor with tractor inadvertently in gear. Tractor rolls while mounting/dismounting. Tractor rolls during hitching/unhitching of equipment.
	Bystander or on-ground helper is run over	Bystander incidents often involve small children the operator does not see. On-ground helper incidents are similar to operator-on-the-ground incidents.
Power take-off (PTO)	Entanglement with PTO stub shaft	Master shield is missing and PTO is left engaged while tractor is running. Operator may be mounting/dismounting from rear of tractor.
Slips and falls	Mounting/ dismounting from tractor	Wet and/or muddy feet, first/last step is high off the ground, difficult to reach handholds, hurrying, facing wrong way when dismounting.

Hazard	Type of incident	How injury occurs
Noise-induced hearing loss	Operating tractor	The tractor muffler may be missing, damaged, or is a non-recommended replacement; tractor engine is not maintained properly; metal weather cab redirects sound back to the operator. Damaging noise level may come from a combination of tractor and attached machine. (Older tractors generally produce louder sounds than newer tractors.)

13.2.1 Overturns

The central concept in tractor stability/instability is centre of gravity (CG). A tractor's CG is the point on the tractor where all parts balance one another. For example, when a two-wheel-drive tractor is sitting with all wheels on level ground, the CG is typically about 25.4 cm above and 0.6 m in front of the rear axle and in the centre of the tractor body. For four-wheel-drive and centre-articulated tractors, the CG is located slightly more forward. For a tractor to stay upright, its CG must stay within the tractor's stability baseline. Stability baselines are essentially imaginary lines drawn between points where tractor tires contact the ground. A tractor's CG as such does not move, but its relationship with stability baselines may change. This most often occurs as the tractor moves out of a perfectly level position, such as onto a slope. A changing relationship between CG and stability baseline means the tractor is moving toward an unstable position. If the CG-stability baseline relationship changes significantly (e.g., the tractor CG moves beyond the stability baseline), the tractor rolls over. If equipment such as a front-end loader, a round bale lifting fork or a chemical side-saddle tank is mounted on the tractor, the additional weight shifts the CG toward that piece of equipment. As mounted equipment is raised, the CG is raised.

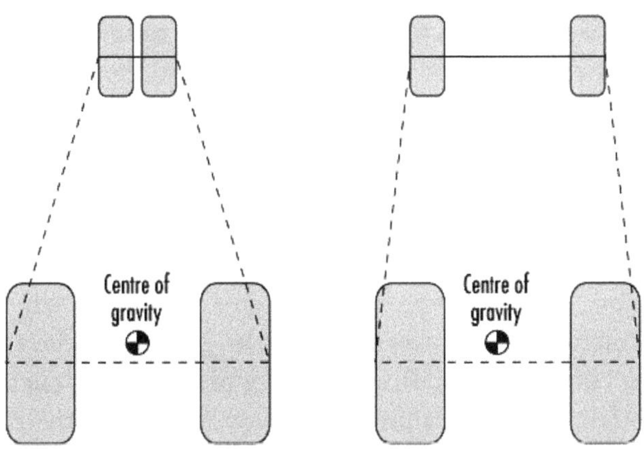

Figure The stability baseline of a tricycle tractor and a wide front-end tractor, respectively

Other factors important to tractor stability/instability include centrifugal force (CF), rear-axle torque (RAT) and drawbar leverage (DBL). Each of these factors works through the CG. Centrifugal force is the outward force nature exerts on objects that move in a circular fashion. Centrifugal force increases both as the turning angle of the tractor becomes sharper (decreases) and as the speed of the tractor increases during a turn. The CF increase is directly proportional to the turning angle of the tractor. For every degree the tractor is turned tighter, there is an equal amount of increased CF. The relationship between CF and tractor speed, however, is not directly proportional. Finding the increase in CF from turning a tractor at a higher speed (assuming the turning radius stays the same) calls for squaring the difference between the two tractor speeds.

RAT involves energy transfer between the tractor engine and the rear axle of a two-wheel-drive tractor. Engaging the clutch results in a twisting force, called torque, to the rear axle. This torque is then transferred to the tractor tyres. Under normal circumstances, the rear axle (and tyres) should rotate, and the tractor will move ahead. In lay terms, the rear axle is said to be rotating about the tractor chassis. If the rear axle should be unable to rotate, the tractor chassis rotates about the axle. This reverse rotation results in the front end of the tractor lifting off the ground until the tractor's CG passes the rear stability baseline. At this point the tractor will continue rearward from its own weight until it crashes into the ground or another obstacle.

DBL is another principle of stability/instability related to rear overturns. When a two-wheel-drive tractor is pulling a load, its rear tyres push against the ground. Simultaneously, the load attached to the tractor is pulling back and down against the forward movement of the tractor. The load is pulling down because it is resting on the earth's surface. This backward and downward pull results in the rear tyres becoming a pivot point, with the load acting as a force trying to tip the tractor rearward. An "angle of pull" is created between the ground's surface and the point of attachment on the tractor. The heavier the load, and the higher the angle of pull, the more leverage the load has to tip the tractor rearward.

13.2.2 Runovers

There are three basic types of tractor runover incidents. One is when a passenger (extra rider) on the tractor falls off the tractor. A second is when the tractor operator falls off the tractor. The third type occurs when a person already on the ground is run over by the tractor. The person already on the ground may be a bystander (e.g., a non-working adult or a small child), a co-worker or the tractor operator. The tractor runover event often involves trailing machinery hitched to the tractor; it may be the trailing machinery that inflicts the injury. Extra rider injury incidents occur because there is no safe location for an extra person on a tractor, yet the practice of taking extra riders is common, as a means of saving time, for convenience, work assistance or baby-sitting. Whether an extra rider can be justified for any reason is strictly in the eye of the beholder. Safety experts and tractor manufacturers strongly recommend against an operator carrying an extra rider for any reason. This advice, however, conflicts with several factors that farmers must face daily. For instance, it is human nature to want to complete work tasks as easily and quickly as possible; different transportation may call for added expenditure of a meagre money supply; other baby-sitting options simply may not exist; and new tractor drivers must be taught how to operate tractors.

Persons already on the ground, usually tractor operators or children, are occasionally run over by tractors and their attached equipment. Tractor operators sometimes try to start their tractor from the ground, instead of from the operator's seat. Most of these incidents occur with older tractors that will start with the tractor in gear, or on newer tractors where the starting interlocks built into the tractor have been by-passed. Small children, usually under the age of five, are sometimes run over by tractors and machinery that is moved around the farmstead. Often, the tractor operator is unaware that the child is even near the equipment. A loud noise, such as the start-up of a tractor, is often attractive to young children and may draw them near. And the practice of allowing extra riders may bring them running to the tractor.

Tractor safety rules include:

- ☆ The most important safety device for a tractor is a rollover protective structure (ROPS). This device, along with a properly buckled seat-belt, prevents an operator from being crushed by the tractor during a rollover.
- ☆ A ROPS enclosed cab provides even more protection, as cabs also provide protection from adverse weather elements and from falling off the tractor.
- ☆ A master shield over the PTO stub shaft protects against PTO entanglement.
- ☆ The one seat–one rider rule and other safe operating practices must be followed.
- ☆ Operator manuals must be read to learn how to safely operate the machine.
- ☆ Workers must be physically, psychologically and physiologically capable of operating a given machine.

Once a potential hazard has been identified, assess the likelihood of an injury or hazardous incident occurring. For example, risk to children playing near a tractor will vary, depending on what the tractor operator is doing, how close they are to the tractor and whether the operator knows they are there. Consider ways of minimizing risk.

- ☆ Read and follow safety procedures in the manufacturer's manual.
- ☆ Ensure an approved cab or rollover protective structure (ROPS) is fitted.
- ☆ Fit and use a seatbelt on tractors with ROPS.
- ☆ If there is a risk from falling objects, fit a fall-on protective structure (FOPS).
- ☆ To reduce risk of back strain, fit a seat with side restraints and a backrest.
- ☆ Wear hearing protection, and remember, not all tractor cabs are sound proof.
- ☆ Keep children away from tractors and machinery.
- ☆ Remove starter keys when tractors are not in use.
- ☆ Have an up-to-date maintenance schedule.
- ☆ Follow safe maintenance and jacking procedures.
- ☆ Ensure the operator is properly trained for each type of tractor work.
- ☆ Always mount and dismount on a tractor's left side - to avoid controls.
- ☆ Adjust the seat so all controls are safely and comfortably reached.
- ☆ Keep all guards in place, including the power take-off (PTO).
- ☆ Operate the self-starter from the operator position only.
- ☆ Never carry passengers.

When Operating a Tractor

- ☆ Drive at speeds slow enough to retain control over unexpected events.
- ☆ Reduce speed before turning or applying brakes.
- ☆ Watch out for ditches, logs, rocks, depressions and embankments.
- ☆ On steep slopes, without a trailed implement, reverse up for greater safety.
- ☆ Engage the clutch gently at all times, especially when going uphill or towing.
- ☆ Use as wide a wheel track as possible on hillsides and sloping ground.
- ☆ Descend slopes cautiously in low gear, using the motor as a brake.
- ☆ Never mount or dismount from a moving tractor
- ☆ Ensure the park brake is on and operating effectively before dismounting.
- ☆ Take short breaks regularly when working long hours.

When Towing Implements

- ☆ Fit attachments according to the manufacturer's instructions.
- ☆ Always attach implements to the draw bar or the mounting points provided by the manufacturer.

- ☆ Never alter, modify or raise the height of the draw bar unless provided for by the manufacturer.
- ☆ Regularly check safety pins on towed lift-wing implements, to ensure they are not worn.
- ☆ Ensure all guards on towed implements are in place before operating.
- ☆ Never hitch above the centerline of the rear axle, around the axle housing or to the top link pin.
- ☆ Never adjust or work on implements while they are in motion.
- ☆ Never attach implements unless the PTO shaft is guarded.
- ☆ When parking, always lower the three-point linkage and towed implement.

To Avoid Strain Injury

- ☆ Adjust the tractor seat for back support and comfort.
- ☆ When buying a tractor, ensure seating is safe and comfortable.
- ☆ Check seat height, seat depth, backrest height and angle, fore and aft movement, seat tilt, firm padding, partial pivoting (if you have to spend long periods looking behind you), and vibration-absorbing suspension.
- ☆ Dismount every hour or so, and spend 5 or 10 minutes doing something active.
- ☆ Plan for your next tractor to include suitably low steps, handgrips, adequate doorway and cab space, and a safe mounting platform.
- ☆ Dismount by climbing down - not jumping down - and use each provided foot and handhold.

13.2.3 Tractor Maintenance

People have been killed and seriously injured doing maintenance and repairs to farm tractors. Major hazards can occur when tractors are jacked and wheels are removed without safe working procedures. These risks are magnified on soil. Regular workshop maintenance of farm tractors and trailed implements can prevent hazardous incidents in the field.

When planning tractor maintenance, check the right equipment is available for safe jacking, removal of wheels and other tasks. People doing the job should be experienced, and there should be agreed safe procedures. Heavy lifting and carrying can cause strain injuries. Children should be kept away from tractor workshops. Field repairs present specific injury risks.

The greater the risk of an injury or a dangerous incident occurring, the more urgent the need for changes to be made to minimize or eliminate the risk.

Here are some ways of improving tractor maintenance safety.

- ☆ Routinely adjust brakes, clutches and drives, according to the manual.
- ☆ Ensure steering, exhaust system and brakes are in top condition.
- ☆ Stop the motor before re-fuelling, servicing or greasing and, if possible, wait until the engine is cold before re-fuelling.
- ☆ Never remove or replace belts while pulleys are under power.
- ☆ Keep steps and working platforms free of grease and oil to avoid slips and falls.
- ☆ If the engine overheats, allow time for it to cool off before removing the radiator cap.

When jacking a Tractor

- ☆ Jack on a flat surface, ideally a concrete floor.
- ☆ Avoid working alone. Ask somebody with training and experience to help.
- ☆ Refer to the manufacturer's manual on safe jacking, or seek professional advice.
- ☆ Where jacking points are not identifiable, jack from the lowest possible point.
- ☆ Use jacks that comply with applicable standards.
- ☆ Use vehicle stands that comply with applicable standards, and are designed for the load to be suspended.

Blocks and Chocks

- ☆ Ensure wooden blocks for jacking are of hardwood, e.g. jarrah or karri, with a surface area that will support the tractor's weight on soft soils.
- ☆ Chock all wheels that will remain on the ground, using big wooden chocks at the front and rear of each wheel. Don't use rocks; they're too unstable.
- ☆ Chock all wheels on articulated vehicles to stop them twisting sideways during jacking.
- ☆ Before jacking, apply brakes, place in gear - or automatic park - and switch ignition off.
- ☆ Stay clear of the tractor while operating the jack.

When removing Wheels

- ☆ Loosen wheel nuts before the wheel is off the ground, to avoid any movement that could dislodge the tractor.
- ☆ Before removing a tractor tire from a rim, release all water and air pressure.
- ☆ To avoid serious injuries, it is recommended that work performed on split rims be done by the professional. Therefore, farm workers should not work on split rims.
- ☆ Never jack more than one wheel off the ground at a time in the field.
- ☆ If both rear wheels have to be removed, work on a flat, level concrete floor, in the workshop.
- ☆ When removing rear wheels, ensure the front wheels are immobilized by fixing wedges between axle and body.

13.3 GRAIN HARVESTING EQUIPMENT SAFETY

Combine operator Safety

Good safety habits are vital for anyone who operates a combine, corn picker or other grain-harvesting machine. Failure to observe safety practices can be fatal!

However, constant alertness is also necessary to prevent machinery accidents--accidents that often happen in spite of machinery that is designed for safety.

Machinery operators are not in top physical or emotional condition when they are tired, ill, worried, angry, or have their minds on something else. Accidents are most likely to happen under these conditions.

The combine operator is responsible not only for his safety but also for the safety of others who may be working on or just be near the machine. The operator must be aware of hazards and remain alert to situations that are potentially dangerous. This includes pre-operational checks, starting, transporting, towing, operating, field repair and maintenance and stopping the combine.

Key points

1. **Hand Signals.** A set of hand signals has been endorsed by several safety institutions. Since spoken instructions are very difficult to hear over the sounds of a combine, knowledge of hand signals can be extremely helpful to the operator when maneuvering a combine, especially in tight places.
2. **Safety Before Starting.**
   a. Before attempting to operate a combine, study the operator's manual. It has information on general safety rules, plus specific safety recommendations for the particular machine. The more you know about the combine, the better prepared you will be to safely operator it.

b. The exhaust fumes from a gasoline or diesel engine are very poisonous. If the combine is run inside a building, be sure to open the doors to provide good ventilation.
c. Always clean the combine before starting. Trash around the exhaust system can cause fires. Oil, grease or mud on ladders or the platform can cause serious falls. If the combine is equipped with a cab, clean the glass to provide maximum visibility.
d. Check the tire pressure each day. Under-inflation can cause buckling of the sidewall, which can cause dangerous tire failure. Over-inflated tire have a great deal of "bounce" and cause upsets more readily that tires with correct pressure.
e. Check the brakes once a week. With hydraulic brakes, make sure that the master cylinder is full of fluid and that no air is present in the lines. Adjust the pedal free travel, if necessary, so that the brakes are engaged with the pedals an equal distance from the floor of the platform. Check the operator's manual for specific instructions.
f. Check the threshing cylinder-rocking bar to see it is clear of the cylinder.
g. Make sure that all shields and covers are in place and fastened securely.
h. Remove or stow all service equipment.
i. Always use the handrails and ladders provided on the combine for safe mounting and dismounting.
j. Be sure that all PTO covers, safety stands and shields are on the machine before taking it to the field.

3. **Starting the Combine.**
   a. Before mounting the combine, make sure that everyone is clear of the machine. Do not allow anyone to ride with you, unless combine is equipped with a passenger seat.
   b. Before starting the combine:
   - Disengage header drive.
   - Disengage separator drive
   - Place gearshift in neutral
   - Depress clutch pedal

   c. Be careful when using diesel starting fluid. It is extremely flammable.
   d. If it is necessary to use jumper cables to start the combine, be careful to avoid sparks around the battery. Hydrogen gas escaping from the battery can explode. Follow the operator's manual instructions for using jumper cables.

4. **Transporting the Combine.**
   a. Always keep your mind on the dangers of driving the combine on public roads. Beside maintaining control of the machine, you must watch for obstacles on the road, pedestrians and traffic.
   b. High speed is the leading cause of accidents. Never drive faster that the road conditions allow for safe operation. Anticipate dangers and slow down to avoid accidents.
   c. Make sure you are familiar with local traffic laws. Check the safety flashers and small moving vehicle (SMV) emblems to be sure they are clean and visible.
   d. Always lock the brake pedals together. If the combine is not equipped with locking mechanism, be sure to depress both pedals at the same time evenly. Applying only one brake, or applying one harder than the other can cause the combine to swerve and perhaps tip over.
   e. Be careful when applying brakes when a header is attached to the combine. The added weight up front can cause the combine to tip forward if the brakes are applied abruptly. Always drive slow enough to allow controlled application of brakes at all times.
   f. Always check headlights and safety flashers to make sure they are properly adjusted and in working order.
   g. Put the unloading auger in the transport position. Be certain it is not blocking a safety flasher or SMV emblem.
   h. On self-propelled combines, never use the header safety support when transporting the machine. Raise the header enough for safe ground clearance, but not high enough to reduce visibility.
   i. On pull-type combines, always use header support when transporting. Towing at transport speeds can be hazardous because of side forces on the tractor when stopping too quickly. Side forces from slowing a combine too quickly may cause a tractor to skid, especially on loose gravel. Slowing down while turning can cause jack-knifing. Slow down before the corner so the towed combine doesn't get out of control.
   j. Watch for low power or telephone lines, bridges, buildings and any other obstacles, to make sure you can pass under them safely. Always keep as far to the right of the roadway as possible. Keep a careful watch to see that you have safe clearance on both sides.
   k. Always sit down when traveling at high speeds or going over rough terrain.
   l. Be careful when making turns. Make sure that the rear of the combine will clear obstacles when it swings around. Avoid sharp turns. Turning too sharply at high speed can cause the machine to turn over.

   m. Because the wheels for steering are in the back, self-propelled machines often fishtail when turned too quickly at transport speeds. Steering to the right will whip the rear to the left, and vice versa. Steering suddenly to the right when meeting oncoming traffic causes the back of the combine to swing out into the path of on coming traffic.
   n. Slowing or braking too rapidly could cause loss of some steering control (weight on rear wheels). This is most noticeable when driving with a corn head or some other heavy header raised high. In this case, most of the weight will be on the drive wheels. Install rear wheel weights. Keep header as low as possible. Use the variable speed drive or engine throttle to slow the machine. Reduce speed before you need to apply brakes and always lock brake pedals together.
   o. Never depress the clutch pedal or take the combine out of gear to coast down hill. When the combine is moving it is impossible to shift the transmission back in gear. Always maintain complete control of the combine. The same applies to tractors that are towing pull-type combines.
5. **Towing the combine.**
   a. If the combine must be transported over long distances, it is safer to haul it on a large truck or a special low trailer.
   b. Never tow the combine at speeds higher than 20 mph.
   c. Always keep the transmission in neutral or in the "tow" position, if the combine is so equipped.
   d. Never tow a combine equipped with hydrostatic drive. Towing can cause damage to the drive unit. Instead, haul the combine.
6. **Operating the combine.**
   a. Never operate the combine if you are ill or sleepy. Operating safety depends on alert, efficient handling of the combine.
   b. Wear safety glasses at all times.
   c. Wear clothing that fits snugly to avoid catching clothing in moving parts.
   d. Never let anyone ride on the combine unless it is equipped with a passenger seat. A rider's clothing may become entangled in moving parts, or he may be thrown off the machine.
   e. Before starting to harvest a field, check it carefully for ditches, fences or other obstacles. Be aware of weather conditions, which present safety hazards.
   f. Be especially careful when operating on hillsides. Avoid sharp turns that could tip the combine over. Beware of ditches or obstacles--they are doubly dangerous on slopes.

g. If grain tank extensions are used, remember that the added weight may make the combine top heavy and more subject to upsets.
h. Never travel over 10 mph (16 km/h) with a full grain tank. The added weight makes the combine more difficult to maneuver and easier to upset.
i. Always sit down when traveling over rough terrain. A sharp jolt can throw you from the platform or away from he controls.
j. Hillside combines are equipped with automatic or manual leveling devices. Hydraulic cylinders act to level these combines on steep slopes. These machines are equipped with a warning signal that indicates when the leveling system has reached its limit. Be especially careful after the device activates.
k. When using the steering brakes, always turn the steering wheel before applying the steering brakes. Failure to do so can cause the combine to swerve and turn dangerously.

7. **Field repair and maintenance safety.**
    a. Always keep the machine clean. Field trash around the exhaust system can cause fires. Mud, grease or oil on the operator's platform or ladders can cause falls.
    b. Before lubricating or adjusting the combine, disengage all drives and stop the engine. Never leave the operator's platform with the engine running.
    c. Make sure that the header drive and separator drive are disengaged before attempting to clean the combine. Never try to unclog the machine with a stick or pole with the machine is running. The stalk rolls on a corn head can pull a 12 foot (3.6 cm) stick through in one second--shorter sticks or stalks even faster--before you can let go.
    d. On a pull-type combines, always disengage the PTO and turn off the tractor before attempting to unclog, adjust or lubricate the machine.
    e. Always stop the machine before opening the inspection doors.
    f. Keep all shields in place. After working on the combine, make sure the shields are fastened securely.
    g. When operating in very dusty or noisy locations, wear goggles and ear plugs to insure safe visibility and prevent hearing loss. Never wear loose clothing that can become entangled in moving parts.
    h. Stay clear of moving parts at all times.
    i. Keep belts and chains properly adjust and aligned.
    j. Don't rely on the hydraulic system for support when working under the machine header. Always use the stops or supports provided on the machine. If no safety device is provided, block the header securely.

k. When adjusting the wheel spacing, make certain the machine is blocked. Never rely on jacks alone for support.
l. Always support the reel arm securely when adjustments are being made.
m. Be careful when removing heavy parts. Make certain they are held firmly to avoid dropping them. Have someone help you with heavy jobs.
n. When operating in dry fields. Install a spark arresting muffler to prevent fire.
o. Avoid sparks or open flames when working the battery. Hydrogen gas escaping from the battery may explode.
p. When possible always refuel the combine outside the field. Let the engine cool before attempting to refuel and never smoke around fuels.
q. Allow the system to cool and remove the radiator cap slowly, turning it until pressure escapes through the overflow pipe. Make sure all pressure is relieved before removing the cap.
r. Stay clear of the exhaust system until it cools.
s. High-pressure fluid leaks in the hydraulic or diesel fuel system are very dangerous. The leaks can be invisible and still have enough pressure to penetrate the skin. When checking for leaks, use a piece of cardboard. If an injury does occur, seek medical aid immediately.
t. Always carry a first aid kit and fire extinguisher on the combine.

8. **Stopping the combine safely.**

   To make sure drive units do not cause injury when the machine is started again, do the following when stopping the combine.

   a. Disengage header drive
   b. Disengage separator drive
   c. Place gearshift lever in neutral
   d. Lower header
   e. Apply parking brake
   f. Remove ignition key to prevent tampering or accidental starting.

Spot The Hazard

Look for hazards relating to:

HANDLING:

☆ Children playing near hay balers, carriers and stackers.

☆ Training of operators handling tractors, front-end loaders or forklifts.

☆ Use of two-poster tractor ROPS for baling - they offer no operator protection from bales falling back off forks or bale-loading frames. Tractors with cabs, FOPS (fall on protective structure) or four-poster ROPS are safer.

- ☆ Makeshift or poorly fitting bale-loading attachments on tractors and forklifts.
- ☆ Carrying bales too high off the ground.
- ☆ Insufficient counterbalance on tractor or forklift vehicle.
- ☆ Hydraulic control valve should be specific to the front-end loader attachment.

Baler Operation:

- ☆ Baler properly connected to the tractor.
- ☆ Adequate safety guards fitted.
- ☆ Nobody allowed to ride on the baler.
- ☆ Prevent others getting too close to the baler.
- ☆ Build-up of loose, combustible material in the baler.
- ☆ Fire extinguisher fitted to the machine.
- ☆ Disengage PTO and apply fly wheel brake prior to making baler adjustments.
- ☆ Stop engine and apply fly wheel brake before repairs or "stringing up" the baler.
- ☆ Extra care and attention when reversing or turning the machine,
- ☆ working at night,
- ☆ loading onto a truck, and unloading.

Stacking:

- ☆ Loading and stacking on uneven ground.
- ☆ Stacks under or near overhead powerlines
- ☆ Stacks of round bales inadequately chocked and border posted.
- ☆ Damaged bales at base of stack – e.g., from vehicles, cattle or rodents.
- ☆ Unstable heights and loose stacking.
- ☆ Bales stacked higher than safe operating height of farm tractor or forklift.
- ☆ Children playing on stacked bales, particularly during stacking or unstacking.
- ☆ Lack of training, experience and protection for people doing hay baling, stacking and loading.
- ☆ Handling more bales than safe for the loader.

Transporting:

- Sturdiness of trailers carrying heavy loads of bales.
- Restraining frames back and front of trailer.
- Hooks fitted so ropes can be used to secure load.
- Roads too close to or below powerlines.
- Rough terrain causing bales to become unstable.
- Safe speeds at all times.
- People riding on loaded hay trailers - highly dangerous.

Assess the Risk

Check each hazard that has been spotted to assess:

- Likelihood - how likely is this hazard to injure someone? and
- Severity - how severe would that injury be?

List all the hay baling and stacking hazards spotted, and number them in order of priority, so that those most likely to cause injury or harm can be tackled first.

Because most large hay bales can kill or seriously injure anyone they fall or roll on, any risk of a hazardous incident should be assessed as requiring urgent attention. And as children are the most vulnerable, consider child injury risks top priority.

Make the Changes

Handling:

- Keep children away from hay baling and stacking operations.
- Make sure operators and handlers are properly trained and physically capable of tasks.
- Consider fall-arrest protection for people working at heights.
- Use tractors with four-poster ROPS, FOPS or cabs for protection against falling bales.
- Replace risky attachments with manufacturer approved attachments.
- Ensure loader or forklift operators transport bales close to the ground.
- Avoid sharp turns and unsafe speeds.
- Make sure vehicle controls are fitted specifically for the attachment in use.

Stacking:

- Stacks should be on firm, level ground, away from fire hazards, sources of ignition, overhead powerlines, dwellings, boundary fences and footpaths.

- ☆ Make sure stack and load heights do not exceed the lifting capabilities of the farm handling equipment.
- ☆ Big bales should be stacked to a maximum of four bales high.
- ☆ High density bales can be stacked up to six layers high.
- ☆ Wherever possible, stack big rectangular, square or high density bales by overlapping, to form a stable stack.
- ☆ Do not use bale lifting equipment to raise people on or off stacks.

Transporting:

- ☆ Do not allow people to ride on stacked trailers.
- ☆ Be aware of overhead obstructions, like trees, bridges and powerlines.
- ☆ Avoid rough ground that could cause bales to dislodge.
- ☆ Ensure loads are adequately secured.
- ☆ Do not overload vehicles beyond legal limits.

13.4 TILLAGE EQUIPMENT

While the skilled operator of tillage equipment avoids errors with very little conscious thought, accident studies show that hurrying and human error are responsible for or are involved in the vast majority of equipment accidents. An operator must have an understanding of the function, operation and limitations of the equipment he/she is operating and the operator must resist the temptation to be hurried into an accident.

Key Points

Moldboard Plows

1. Provide adequate front-end weight for tractor stability in transport and operation, particularly with integral and semi-integral plows. Never pull from any point higher on the tractor than the recommended hitch point.
2. Use extreme caution and reduce speed when transporting the plow and the tractor over rough ground.
3. Avoid sharp turns at high speeds, especially on slopes.
4. On tight turns, avoid swinging rear of plow into fences or other obstacles.
5. Turning stops on semi-integral plows limit turning radius. Shorter turns may severely damage plow frame and tractor hitch.
6. Never carry passengers on the tractor or permit others to ride on the plow -- particularly plows with automatic reset.
7. Always lower the plow when not in use or left unattended.

8. Lower the plow and securely pin the parking stand before detaching the plow from the tractor.
9. Always use proper lighting, reflectors, slow moving vehicle (SMV) emblem, and other safety devices for road travel as required by state and local laws.
10. When hitching drawn plows, always use a hitch pin with adequate strength for the tractor-plow combination.

Disk Plows

1. Integral plows are transported completely raised, and all weight is carried by the tractor 3-point hitch. Adequate tractor front-end weights are required to offset the plow weight.
2. When the transporting on a road or highway, always display SMV emblem and use lights and reflectors as required by state and local regulations.
3. Semi-integral plows are quite long and caution must be used when turning to prevent swinging the plow into fences or irrigation ditches.
4. Reduce speed when transporting over rough ground, and avoid quick, sharp turns at high speeds.
5. When transporting semi-integral or drawn plows, always install cylinder locks to prevent accidental lowering of the plow. Relieve the load on hydraulic cylinders before starting to transport.
6. Lower the plow to the ground or install hydraulic cylinder locks when the plow is not in use.
7. Watch for other people when raising, lowering, or indexing the plow.
8. Never permit anyone to ride on the plow, and allow only the driver on the tractor. Do not permit children to play on or near the plow either when parked or in operation.
9. Lower the parking stand and securely pin it in place before detaching integral or semi-integral plows from the tractor.

Disk Tillers or Oneways

1. Keep SMV emblem clean and prominently displayed. Do the same with reflectors and warning lights as required by state and local regulations.
2. Never allow anyone but the operator to ride on the tractor.
3. Never ride or permit others to ride on the tiller.
4. Lower the tiller to the ground when not in use.
5. Secure the machine in the raised position by installing safety locks or hold-up pins when servicing or cleaning it.
6. Disk blades are extremely sharp; be very careful when working or making adjustments in the disk area.

7. Never walk close beside the rear wheel when the tiller is in operation. A sudden imbalance of forces could cause this wheel suddenly to jump to the left.
8. Never grease, oil, or adjust the tiller while it is in operation.
9. Escaping hydraulic oil under pressure can cause serious personal injury and infection. Therefore, be sure all connections are tight and that oil lines are undamaged. Always relieve hydraulic pressure in lines before disconnecting hoses.

Chisel Plow

1. Reduce speed when transporting chisel plows over rough or uneven terrain.
2. Use lock-up straps or transport locks when transporting a chisel plow.
3. Be sure wings are locked in the folded position before traveling.
4. Use proper lights, reflectors, and a clean SMV emblem when transporting equipment on road or highway.
5. The transport width of most folding-section plows exceeds maximum width of normal vehicles. Therefore, use extreme caution when meeting other traffic, to avoid collisions and the possibility of transport wheels dropping into holes, drains, or ditches along the road edge.
6. Allow only the operator to ride on the tractor.
7. Never allow passengers to ride on the chisel plow.
8. Never allow anyone to stand or work near the chisel plow when it is in operation, particularly when raising or lowering outriggers.
9. Do not permit children to play on or near the chisel plow during operation or storage.
10. Provide adequate tractor front-end ballast for stability in transport and operation, especially with integral models.
11. Be particularly careful of escaping hydraulic fluid, which can penetrate the skin and cause serious infection or reaction if not given immediate medical treatment.
12. Do not unhitch from the tractor or store a chisel plow when outriggers are in the raised position.
13. Make sure raised outriggers will safely pass under power and telephone lines.

Disk Harrows and Offset Disks

1. Always lock safety lock during transport, if the disk is to be left raised for an extended period, or while working on the machine.
2. Never depend on tractor hydraulic pressure to carry harrow weight in transport--use safety lock, and relieve pressure in cylinders.

3. Lower integral harrows to ground each time tractor engine is shut off, and any time harrow is being serviced or repaired. If it must be raised for repairs, securely block the frame to prevent accidental lowering.
4. Always use lights, reflectors, and SMV emblem when transporting, day or night.
5. Lock the tractor drawbar in fixed position when transporting wheeled disks.
6. Never transport a disk harrow on its own wheels at more than normal tractor speed, and considerably less than that on rough or uneven ground.
7. Never clean, adjust, or lubricate the harrow while it is in motion.
8. Wear protective gloves when working with or near disk blades.
9. Hydraulic fluid escaping under pressure can penetrate the skin and cause serious infection or reactions. Never use hands to locate the source of a small leak which may be nearly invisible. Obtain immediate medical attention if injured by escaping hydraulic fluid.
10. Park or block the harrow so it cannot roll when unhitched.
11. Make sure wings are securely locked in transport position before moving the harrow.
12. Large disk harrows exceed normal vehicle width, so be particularly careful to avoid collisions when meeting other vehicles on the road. Avoid dropping wheels of tractor or harrow into holes, drains, or ditches along the road.
13. Provide adequate tractor ballast for front-end stability and to prevent excessive slippage.
14. Never allow anyone to ride on tractor drawbar or harrow in operation or transport.
15. Never allow anyone but the operator to ride on the tractor.
16. Lower the machine or install safety lock when storing a disk harrow.
17. Never permit children to play on or near a disk harrow while it is in operation, transport, or storage.
18. Stand clear of harrow wings during folding or unfolding.
19. Remove spring-loaded scrapers in proper order to avoid personal injury. Use care in relieving any spring under tension or compression.
20. Do not make sharp turns with blades down.

Field Cultivators

1. Never exceed recommended transport speed for the cultivator used. If speed is not stated, do not exceed maximum tractor speed.
2. Reduce speed for turning and travel over rough or uneven ground.
3. Use transport locks and relieve pressure in cylinders when transporting field cultivators. Do not depend on hydraulic pressure to carry the weight. Always lock wings in transport position and relieve pressure in cylinders.

4. Never walk or work under wings when they are in the folded position.
5. Follow state and local regulations regarding lights, reflectors, SMV emblem, and maximum width when transporting on roads or highways.
6. Transport width of most field cultivators exceeds normal vehicle width. Therefore, use extreme caution when meeting other vehicles and avoid the possibility of dropping tractor or- implement wheels into holes, drains, or ditches along the road edge.
7. Never permit anyone to ride on the tractor drawbar or cultivator in transport or operation, or to stand near the machine while it is operating--particularly when raising or lowering wings.

Toothed Harrows

1. Provide sufficient front-end weight for safe, stable operation and transport.
2. Use widest practical wheel tread to improve tractor stability, especially when working on steep slopes.
3. Reduce implement to narrowest possible width for transport on roads or highways.
4. Lock wings or folded sections securely in place before transporting equipment.
5. Use lights, reflectors, and SMV emblem as required by law when transporting equipment--day or night.
6. Schedule moves for least hazardous periods; avoid transporting equipment on busy roads, during peak traffic periods, or after dark.
7. Never transport wheeled harrows, harrow carts, or field conditioners at more than tractor transport speeds; transport considerably slower on rough or uneven terrain.
8. Never make sharp turns at high speeds.
9. Never allow anyone to ride on the tractor but the operator.
10. Never allow anyone to ride on the tractor drawbar or implement in operation or transport.
11. Always stop implement and tractor engine to adjust, repair, or lubricate.
12. Lower implements to the ground before stopping tractor engine, before servicing or repairing equipment, or at any time the machines are left unattended.
13. Never make extremely short turns with drag-type eveners which could foul tractor tires.
14. Never park implements where they could be hidden by tall or growing crops, grass, or weeds.

15. Keep PTO shaft properly shielded. Never get off tractor without disengaging PTO and stopping engine.

Packers

1. Always use reflectors, lights, and SMV emblem as required when transporting equipment--day or night.
2. Provide adequate front-end weight for tractor stability in operation and transport of integral harrows. Use maximum allowable front ballast if operating integral harrows in lower gears.
3. Never exceed normal tractor speed when transporting drawn roller harrows, and drive considerably slower than that on rough or uneven ground.
4. Do not transport roller packers over hard-surfaced roads--use carriers.
5. Never permit anyone but the driver to ride on the tractor.
6. Never permit anyone to ride on the tractor drawbar or implement during operation or transport.
7. Install transport lock pin before storing, transporting, or parking drawn implement; do not depend on hydraulic pressure to support the weight. Lower machines to the ground whenever the tractor engine is shut off.
8. Pin tractor drawbar in center before transporting.
9. Never lubricate, adjust, or repair the implement while it is in motion or the tractor engine is running.
10. Always raise spring teeth before lowering roller harrow to the ground for parking or storage.
11. Park or block the implement to prevent rolling when it is disconnected from the tractor.
12. Never try to lift or support the roller harrow on the spring teeth for service or repairs.

Undercutters

1. Use the SMV emblem, lights, and reflectors as required by law for transporting equipment on roads or highways.
2. Install the transport safety lock and relieve hydraulic pressure in the cylinders when transporting.
3. Be certain wings are securely locked in the folded position before transporting.
4. Limit transport speed as recommended--15 miles an hour (24 Km/h) for some machines, even less on rough or uneven terrain.
5. Never ride or allow others to ride on the machine during operation or transport. Allow only the driver on the tractor.
6. Do not unhitch from the tractor or store the machine with wings in the folded transport position.

7. Sweeps are sharp--watch out for them when wings are folded for transport or when servicing, adjusting, or repairing the plow. Always position wrenches to pull away from sharp edges or corners.
8. Never stand with feet under blades while making adjustments or during maintenance. Be extremely careful while working within the implement frame.
9. Do not stand or walk on the plow frame, or under wings when they are folded for transport.
10. Stand with both feet on the same side of the tongue when hitching or unhitching.
11. Do not allow children to play on or near the plow.

Rod Weeders

1. Allow only the operator to ride on the tractor during operation and transport.
2. Never permit anyone to ride on the rod weeder.
3. Never exceed recommended transport speed or, if not stated, maximum tractor speed. Reduce speed on rough or uneven terrain or when turning.
4. Use lights, reflectors, and SMV emblem as required by state and local regulations when transporting equipment, day or night.
5. Avoid busy highways and peak traffic periods if possible. Move equipment only in daytime.
6. Never attempt to repair, adjust, or lubricate the rod weeder while it is in motion.
7. Shut off the engine and relieve the hydraulic pressure in hoses before disconnecting them.
8. Never exceed recommended transport speed, or tractor road speed if maximum is not stated. Reduce speed when turning or crossing rough areas and slopes.
9. Always lower parking stands on integral equipment before detaching from tractor.
10. Always lower equipment or install transport lock before servicing, lubricating or repairing equipment, and when the machine will be left unattended.

INDEX

www.ingramcontent.com/pod-product-compliance
Ingram Content Group UK Ltd.
Pitfield, Milton Keynes, MK11 3LW, UK
UKHW021327190726
13845UKWH00004B/66

9 789349 964327